Cost Optimization of Structures

Cost Optimization of Structures

Fuzzy Logic, Genetic Algorithms, and Parallel Computing

Hojjat Adeli
The Ohio State University, USA

Kamal C. Sarma
Barr Engineering, Inc., USA

John Wiley & Sons, Ltd

Other Wiley Editorial Offices

John Wiley & Sons Inc., 111 River Street, Hoboken, NJ 07030, USA

Jossey-Bass, 989 Market Street, San Francisco, CA 94103-1741, USA

Wiley-VCH Verlag GmbH, Boschstr. 12, D-69469 Weinheim, Germany

John Wiley & Sons Australia Ltd, 42 McDougall Street, Milton, Queensland 4064, Australia

John Wiley & Sons (Asia) Pte Ltd, 2 Clementi Loop #02-01, Jin Xing Distripark, Singapore 129809

John Wiley & Sons Canada Ltd, 6045 Freemont Blvd, Mississauga, ONT, L5R 4J3

Wiley also publishes its books in a variety of electronic formats. Some content that appears in print may not be available in electronic books.

Library of Congress Cataloging in Publication Data

Adeli, Hojjat, 1950–
    Cost optimization of structures: fuzzy logic, genetic algorithms, and parallel computing / Hojjat Adeli, Kamal Sarma.
        p.   cm.
    Includes bibliographical references and index.
    ISBN-13 978-0-470-86733-4
    ISBN-10 0-470-86733-7 (cloth: alk. paper)
    1. Structural optimization—Mathematics.   2. Skyscrapers—Design and construction—Cost control.
    I. Sarma, Kamal.   II. Title.
    TA658.8.A345 2006
    721′.042—dc22

British Library Cataloguing in Publication Data

A catalogue record for this book is available from the British Library

ISBN-13 978-0-470-86733-4 (HB)
ISBN-10 0-470-86733-7 (HB)

Typeset in 10.5/13pt Times by Integra Software Services Pvt. Ltd, Pondicherry, India
Printed and bound in Great Britain by TJ International, Padstow, Cornwall
This book is printed on acid-free paper responsibly manufactured from sustainable forestry in which at least two trees are planted for each one used for paper production.

Dedicated to

Djafar Adeli and Mokarrameh Soofi

Vikram, Lona, and Sharmistha Sarma

Contents

Preface   xi

Acknowledgments   xiii

About the Authors   xv

1 Introduction   **1**
1.1 The Case for Cost Optimization   1
1.2 Cost Optimization of Concrete Structures   2
     1.2.1 Concrete Beams and Slabs   3
     1.2.2 Concrete Columns   11
     1.2.3 Concrete Frame Structures   12
     1.2.4 Bridge Structures   14
     1.2.5 Water Tanks   16
     1.2.6 Folded Plates and Shear Walls   17
     1.2.7 Concrete Pipes   17
     1.2.8 Concrete Tensile Members   17
     1.2.9 Cost Optimization Using the Reliability Theory   18
     1.2.10 Concluding Comments   20
1.3 Cost Optimization of Steel Structures   20
     1.3.1 Deterministic Cost Optimization   20
     1.3.2 Cost Optimization Using the Reliability Theory   31
     1.3.3 Fuzzy Optimization   33
     1.3.4 Concluding Comments   35

2 Evolutionary Computing and the Genetic Algorithm   **37**
2.1 Overview and Basic Operations   37
2.2 Coding and Decoding   39

2.3 Basic Operations in Genetic Algorithms   40
2.4 GA with the Penalty Function Method   43
     2.4.1 Problem Formulation for Axial Force (Truss)
            Structures   43
     2.4.2 Genetic Algorithm with the Penalty Function
            Method   45
2.5 Augmented Lagrangian Method   47
2.6 GA with the Augmented Lagrangian Method   48
     2.6.1 Problem Formulation for Axial Force (Truss)
            Structures   48
     2.6.2 Genetic Algorithm with the Augmented Lagrangian
            Method   49

3 Cost Optimization of Composite Floors   53
3.1 Introduction   53
3.2 Minimum Cost Design of Composite Beams   54
     3.2.1 Cost Function   54
     3.2.2 Constraints   55
     3.2.3 Problem Formulation as a Mixed Integer–Discrete
            Nonlinear Programming Problem   61
3.3 Solution by the Floating-Point Genetic Algorithm   62
     3.3.1 Binary Versus Floating-Point GA   62
     3.3.2 Crossover Operation for the Floating-Point GA   62
     3.3.3 Mutation Operation for the Floating-Point GA   63
     3.3.4 Floating-Point GA for Cost Optimization
            of Composite Floors   63
3.4 Solution by the Neural Dynamics Method   65
3.5 Counter Propagation Neural (CPN) Network for Function
     Approximations   68
3.6 Examples   71
     3.6.1 Example 1   71
     3.6.2 Example 2   72

4 Fuzzy Genetic Algorithm for Optimization of Steel Structures   77
4.1 Introduction   77
4.2 Fuzzy Set Theory and Structural Optimization   79
4.3 Minimum Weight Design of Axially Loaded Space
     Structures   82
4.4 Fuzzy Membership Functions   85
4.5 Fuzzy Augmented Lagrangian Genetic Algorithm   87

4.6   Implementation and Examples   92
      4.6.1   Example 1   93
      4.6.2   Example 2   93
4.7   Conclusion   98

5 Fuzzy Discrete Multi-criteria Cost Optimization of Steel
   Structures   101
5.1   Cost of a Steel Structure   101
5.2   Primary Contributing Factors to the Cost of a Steel
      Structure   102
5.3   Fuzzy Discrete Multi-criteria Cost Optimization   105
5.4   Membership Functions   110
      5.4.1   Membership Function for Minimum Cost   110
      5.4.2   Membership Function for Minimum Weight   110
      5.4.3   Membership Function for Minimum Number of
              Section Types   110
5.5   Fuzzy Membership Functions for Criteria with Unequal
      Importance   112
5.6   Pareto Optimality   112
5.7   Selection of Commercially Available Discrete Shapes   114
5.8   Implementation and a Parametric Study   117
5.9   Application to High-Rise Steel Structures   118
      5.9.1   Example 1   118
      5.9.2   Example 2   119
5.10 Concluding Comments   123

6 Parallel Computing   125
6.1   Multiprocessor Computing Environment   125
6.2   Parallel Processing Implementation Environment   128
      6.2.1   OpenMP Data Parallel Application Programming
              Interface (API)   128
      6.2.2   Message Passing Interface (MPI)   130
6.3   Performance Optimization of Parallel Programs   130

7 Parallel Fuzzy Genetic Algorithms for Cost Optimization of
   Large Steel Structures   133
7.1   Genetic Algorithm and Parallel Processing   133
7.2   Cost Optimization of Moment-Resisting Steel Space
      Structures   135

7.3   Data Parallel Fuzzy Genetic Algorithm for Optimization
      of Steel Structures Using OpenMP   136
7.4   Distributed Parallel Fuzzy Genetic Algorithm for
      Optimization of Steel Structures Using MPI   138
      7.4.1   Processor Farming Scheme   138
      7.4.2   Migration Scheme   140
7.5   Bilevel Parallel Fuzzy GA for Optimization of Steel
      Structures Using OpenMP and MPI   142
      7.5.1   Bilevel Parallel Fuzzy GA with the Processor
              Farming Scheme   145
      7.5.2   Bilevel Parallel Fuzzy GA with the Migration Scheme   146
7.6   Application to High-Rise Building Steel Structures   147
      7.6.1   Example 1   147
      7.6.2   Example 2   149
7.7   Parallel Processing Performance Evaluation   155
      7.7.1   Data Parallel Fuzzy GA Using OpenMP   155
      7.7.2   Distributed Parallel Fuzzy GA Using MPI   157
      7.7.3   Bilevel Parallel Fuzzy GA Using OpenMP and MPI   160
7.8   Concluding Comments   164

8 **Life-Cycle Cost Optimization of Steel Structures**   **165**
8.1   Introduction   165
8.2   Life-Cycle Cost of a Steel Structure and the Primary
      Contributing Factors   167
8.3   Formulation of the Total Life-Cycle Cost   170
8.4   Fuzzy Discrete Multi-criteria Life-Cycle Cost Optimization   171
8.5   Application to a High-Rise Building Steel Structure   174

Appendix A   **177**

Appendix B   **181**

References   **185**

Index   **201**

Preface

During the past 45 years a significant amount of research has been published in the area of structural optimization. The great majority of these papers deal with the minimization of the weight of the structure. While weight of a structure constitutes a significant part of the cost, a minimum weight design is not necessarily the minimum cost design. Only a small fraction of the papers published on structural optimization deal with the cost optimization problem. Most of these papers deal with structural elements such as beams. Few journal papers have been published on cost optimization of realistic three-dimensional structures. As such, there is a need to perform research on cost optimization of realistic three-dimensional structures, especially large structures with hundreds of members where optimization can result in substantial savings. The results of such research efforts will be of great value to practicing engineers.

Another important reason for advocating and advancing the use of optimization technology in the design of structures is automating the complicated process of structural design. In the ground-breaking book by H. Adeli and H. S. Park, *Neurocomputing for Design Automation*, CRC Press, Boca Raton, Florida, 1998, a patented model was presented for fully automated design of very large structures and used for the fully automated design of a 144-story superhigh-rise steel modified tube-in-tube structure with over 20 000 members subjected to the actual constraints of commonly used design code, the American Institute of Steel Construction Allowable Stress Design (ASD) and Load and Resistance Factor Design (LRFD) codes. Today, even the most advanced commercial design software systems do not provide a fully automated system. Fully automated structural design and cost optimization are where the large-scale design technology should be heading. As the first book on *cost optimization of structures* we hope that this book will guide the profession in that direction.

Acknowledgments

Parts of the work presented in this book were published by the authors in a number of journal articles in *Journal of Structural Engineering,* and *Practice Periodical on Structural Design and Construction* (American Society of Civil Engineers), *Computer-Aided Civil and Infrastructure Engineering* (Blackwell Publishers), *International Journal for Numerical Methods in Engineering* (John Wiley & Sons, Ltd), and *International Journal of Space Structures* (Multi-Science Publishing Company) as noted in the list of references and acknowledged in various chapters. Chapter 2 is based on two journal articles by the senior author and his graduate student N.T. Cheng, and Chapter 3 is based on two journal articles by the senior author and his graduate student and research associate Dr H. Kim and are reproduced by permission of the publishers of these journals as noted in the footnotes to these chapters.

The junior author expresses his sincere gratitude and appreciation to Mr James E. Prevost of Barr & Prevost for his support and understanding in completing this work.

About the Authors

Hojjat Adeli is Professor of Civil and Environmental Engineering and Geodetic Science and Lichtenstein Professor in Infrastructure Engineering at The Ohio State University. A contributor to 70 different scholarly journals, he has authored over 400 research and scientific publications in diverse areas of engineering, computer science, and applied mathematics since he received his PhD from Stanford University in 1976 at the age of 26. He has authored/co-authored eleven pioneering books such as *Machine Learning – Neural Networks, Genetic Algorithms, and Fuzzy Systems*, John Wiley & Sons, Ltd, 1995; *Neurocomputing for Design Automation*, CRC Press, 1998; *Distributed Computer-Aided Engineering*, CRC Press, 1999; *High-Performance Computing in Structural Engineering*, CRC Press, 1999; *Control, Optimization, and Smart Structures – High-Performance Bridges and Buildings of the Future*, John Wiley & Sons, Ltd, 1999; *Construction Scheduling, Cost Optimization, and Management – A New Model Based on Neurocomputing and Object Technologies*, Spon Press, 2001; and *Wavelets in Intelligent Transportation Systems*, John Wiley & Sons, Ltd, 2005. He has also edited 12 books including *Intelligent Information Systems*, IEEE Computer Society, 1997.

He is the Editor-in-Chief of two research journals, *Computer-Aided Civil and Infrastructure Engineering*, which he founded in 1986, and *Integrated Computer-Aided Engineering*, which he founded in 1993. He has been a Keynote/Plenary Lecturer at 56 international conferences held in 33 different countries and has served on the advisory/editorial/organizing board of 262 national and international conferences held in 55 different countries. On September 29, 1998, he was awarded a patent for a 'Method and apparatus for efficient design automation and optimization, and structures produced thereby' (United States Patent Number (5,815,394) (with a former PhD student).

He is the recipient of numerous academic, research, and leadership awards, honors, and recognition. In 1998, he was awarded the University

Distinguished Scholar Award by The Ohio State University 'in recognition of extraordinary accomplishment in research and scholarship' and the Senate of the General Assembly of the State of Ohio passed a resolution honoring him as an 'Outstanding Ohioan'. He is the quadruple winner of The Ohio State University College of Engineering Lumley Outstanding Research Award. In 2005 he was elected an Honorary Member of the American Society of Civil Engineers 'for wide-ranging, exceptional, and pioneering contributions to computing in many civil engineering disciplines and extraordinary leadership in advancing the use of computing and information technologies in civil engineering throughout the world'. In 2006 he was awarded The American Society of Civil Engineers' Construction Management Award 'for development of ingenious computational and mathematical models in the areas of construction scheduling, resource scheduling, and cost estimation'.

His research has been sponsored by 20 different organizations including government funding agencies such as the National Science Foundation, US Air Force Flight Dynamics Laboratory, and US Army Construction Engineering Research Laboratory, Federal Highway Administration, state funding agencies such as the Ohio Department of Transportation, Ohio Department of Development, and the State of Ohio Research Challenge Program, professional societies such as the American Iron and Steel Institute and the American Institute of Steel Construction, and corporations such as Cray Research Inc., US Steel, and Bethlehem Steel Corporation.

Kamal C. Sarma is a Senior Bridge Engineer at Barr & Prevost in Columbus, Ohio. He is a registered Professional Engineer in the state of Ohio. He has more than 25 years of work experience in Civil and Structural Engineering and has designed numerous multi-span highway bridges in the state of Ohio. He obtained his Bachelor of Engineering degree in 1976 from Jorhat Engineering College in India and obtained his Master of Structural Engineering degree from University of Roorkee, India in 1984. He received his PhD in Civil Engineering from The Ohio State University in 2001. He worked as a Lecturer and an Assistant Professor in Assam Engineering College for 12 years where he taught courses on design of reinforced concrete and steel structures. As an expert consultant, the Government of Assam selected him as a member of a three-member committee assigned to investigate the development of cracks in the foundation of a thermal power station in Chandrapur, Assam, India. In the United States he also worked as a Senior Software Development Engineer for Qwest Communications and a Field Engineer for K&S Engineers in Highland, Indiana. He performed construction inspection of several multi-storied structures in the Chicago area. He has also consulted in the areas of geotechnical and foundation engineering including slurry wall construction. He is the co-author of 10 research articles in the areas of structural optimization, genetic algorithms, fuzzy systems, and high-performance computing including parallel processing, published in several international research journals.

1

Introduction

1.1 The Case for Cost Optimization

A great majority of structural optimization papers deal with the minimization of the weight of the structure (Vanderplaats, 1984; Arora, 1989; Adeli and Kamal, 1993; Adeli, 1994 – to mention a few). While weight of a structure constitutes a significant part of the cost, minimization of the cost is the final objective for optimum use of available resources. For concrete structures the optimization problem has to be formulated as a cost minimization problem because different materials are involved. In contrast, for steel structures the optimization problem can be formulated as a weight minimization problem. Only a small fraction of hundreds of papers published on optimization of steel structures deals with cost optimization; the great majority deal with minimization of the weight of the structure. In reality, a minimum weight design may not be a minimum cost design. Besides material cost there are many other factors that influence the total construction cost of a structure. Up to the late 1990s, little research work had been reported on the optimization of the overall cost of a three-dimensional steel structure subjected to the constraints of the commonly used design specifications such as AISC ASD and LRFD specifications (AISC, 1995, 2001).

Ideally, the optimization problem should be formulated in terms of the life-cycle cost, which includes the costs of materials, fabrication, erection, maintenance, and disassembling the structure at the end of its life cycle. Some methodologies for determining life-cycle costs and decision making are discussed by Wilson *et al.* (1997). However, prior to the authors' research, little

Cost Optimization of Structures: Fuzzy Logic, Genetic Algorithms, and Parallel Computing H. Adeli and K. C. Sarma © 2006 John Wiley & Sons, Ltd

work on cost optimization of steel structures was reported in the literature. Optimization of total cost as well as the life-cycle cost are very important from the economic point of view and should be the prime focus of structural optimization in the new millennium.

In the traditional optimization algorithms, constraints are satisfied within a tolerance defined by a crisp number. In actual engineering practice constraint evaluation involves many sources of imprecision and approximation. When an optimization algorithm is forced to satisfy the design constraints exactly it can miss the global optimum solution within the confines of commonly acceptable approximations. By taking into account the fuzziness and imprecision in the constraints (the input of the optimization problem) and employing the fuzzy set theory of Zadeh (1965, 1978), one can further reduce the objective function (the output of the optimization problem) and increase the probability of finding the actual global optimum solution substantially.

For the structural optimization methodology in general, and the cost optimization approach in particular, to be embraced by the structural engineering community, the focus of future research should be on large structures subjected to the actual constraints of a commonly used design code such as the AISC ASD (AISC, 1995) or the AISC LRFD (AISC, 2001) code. The true benefit of optimization is realized for large structures with hundreds of members.

In cost minimization additional difficulties are encountered. They include the definition of the cost function and uncertainties and fuzziness involved in determining the cost parameters. As a result, only a small fraction of the structural optimization papers published deal with the minimization of the cost. In this chapter, a chronological review of papers is presented on the cost optimization of concrete and steel structures published in archival journals.

1.2 Cost Optimization of Concrete Structures

Hundreds of papers have been published on optimization of structures during the past four decades. However, only a small fraction of them deal with cost optimization of structures. The great majority of the structural optimization papers are concerned with minimization of the weight of the structure. For concrete structures the objective function to be minimized should be the cost since they are made of more than one material. A review of articles on cost optimization of concrete structures published in archival journals is presented in this section, where interesting and important results and conclusions are summarized. Most of these papers deal with structural elements such as beams. Few journal papers are found on cost optimization of rather realistic three-dimensional structures. As such, there is a need to perform research on cost optimization of realistic three-dimensional structures, especially large structures with hundreds of members where optimization can result in substantial savings. The results of such research efforts will be of great value to practicing engineers.

Concrete structures include reinforced concrete (RC), prestressed concrete, and fiber-reinforced concrete structures. In concrete structures at least three different cost items should be considered in optimization: costs of concrete, steel, and the formwork. This review is presented in terms of different types of concrete structures. Reliability-based cost optimization is reviewed in the last section.

1.2.1 Concrete Beams and Slabs

Most of the papers published on cost optimization of concrete structures are about beams or girders. The general cost function for reinforced, fiber, or prestressed concrete beams can be expressed in the following form:

$$C_m = C_{cb} + C_{sb} + C_{pb} + C_{fb} + C_{sbv} + C_{fib} \tag{1.1}$$

where C_m is the total material cost, C_{cb} is the cost of concrete in the beam, C_{sb} is the cost of reinforcing steel, C_{pb} is cost of prestressing steel, C_{fb} is the cost of the formwork, C_{sbv} is the cost of shear steel, and C_{fib} is the cost of fiber in concrete. For a pretensioned beam equation (1.1) can be written as

$$C_m = w_c L_b \left(A_{cb} - A_{sb} - A'_{sb} - A_{pb} \right) c_c + w_s L_b \left(A_{sb} + A'_{sb} \right) c_s$$
$$+ w_p L_b A_{pb} c_p + L_b p_{fb} c_f + C_{sbv} + C_{fib} \tag{1.2}$$

where L_b is the length of the beam; w, A, c are unit weights, cross-sectional areas, and unit costs, respectively; subscripts b, c, s, p, and f refer to beam, concrete, steel, prestressing, and formwork, respectively; the prime indicates compression steel; and p_{fb} is the cross-sectional perimeter of the form. Equations (1.1) and (1.2) can be reduced for special cases. For example, in the case of an RC beam with no prestressing and fiber, the quantities A_{pb} and C_{fib} are set to zero in equation (1.2).

Goble and Lapay (1971) minimize the cost of post-tensioned prestressed concrete T-section beams based on the ACI code (ACI, 1963) by using the gradient projection method (Arora, 1989). The cost function includes the first four terms in equation (1.1). They state that the optimum design seems to be unaffected by the changes in the cost coefficients. However, subsequent researchers to be discussed later rebut this conclusion.

Kirsch (1972) presents the minimum cost design of continuous two-span prestressed concrete beams subjected to constraints on the stresses, prestressing force, and the vertical coordinates of the tendon by linearizing the nonlinear optimization problem approximately and solving the reduced linear problem by the linear programming (LP) method. His cost function includes only the first and third terms in equation (1.1). Kirsch (1973) extends this work to prestressed concrete slabs.

Friel (1974) finds closed-form solutions for the optimum ratio of steel to concrete for minimum cost simply supported rectangular RC beams using the ultimate moment constraints of the ACI code (ACI, 1971). The cost function is similar to equation (1.1), but neglecting the costs of prestressing steel (C_{pb}) and fiber (C_{fib}) and adding an additional term for increasing the cost due to the increase in the building height. The author concludes that the costs of the formwork and the increase in the height do not influence the optimum cost significantly.

Brown (1975) presents an iterative method for minimum cost selection of the thickness of simply supported uniformly loaded one-way slabs using only the flexural constraints of the ACI code (ACI, 1971). The cost function includes only the first two terms in equation (1.1). The author reports cost savings of up to 17 %.

Naaman (1976) compares minimum cost designs with minimum weight designs for simply supported prestressed rectangular beams and one-way slabs based on the ACI code (ACI, 1971). The cost function includes the first, third, and fourth terms in equation (1.1) and is optimized by a direct search technique (Siddall, 1972). He concludes that the minimum weight and minimum cost solutions give approximately similar results only when the ratio of cost of concrete per cubic yard to the cost of prestressing steel per pound is more than 60. Otherwise, the minimum cost approach yields a more economical solution, and for ratios much smaller than 60 the cost optimization approach yields substantially more economical solutions. He also points out that for most projects in the US the aforementioned ratio is less than 60.

Chou (1977) uses the Lagrange multiplier method for the minimum cost design of a singly reinforced T-beam using the ACI code (ACI, 1971). The author defines only two design variables: effective depth and area of steel reinforcement. The cost function includes the first two terms in equation (1.1).

In the formulation it is assumed that the neutral axis is located inside the flange of the T-section. The author reports a cost reduction up to 14 % of the cost of the beams with a maximum steel ratio.

Gunaratnam and Sivakumaran (1978) present the minimum cost design of RC slabs satisfying the limit state requirements of the British code (CP110, 1972) for members having uniform, triangular, or parabolic moment distribution using a combination of the Lagrange multiplier and graphical methods. Their cost function includes only the first two terms in equation (1.1). They present curves for optimum design parameters as a function of the thickness of the slab. They point out the significant influence of the serviceability limit state of deflection on the optimum design parameters.

Kirsch (1983) presents a simplified three-level iterative procedure for cost optimization of multi-span continuous RC beams with rectangular cross-sections using a cost function consisting of only the first two terms in equation (1.1). In the first level the amount of the reinforcement is found in each critical section for given concrete dimensions and design moments. In the second level the concrete dimensions of each element are found. In the third level design moments are optimized. The author does not, however, consider the constraints of a concrete design code used in practice.

Cohn and MacRae (1984a) consider the minimum cost design of simply supported RC and partially or fully pre-tensioned and post-tensioned concrete beams of fixed cross-sectional geometry subjected to serviceability and ultimate limit state constraints including constraints on flexural strength, deflection, ductility, fatigue, cracking, and minimum reinforcement, based on the ACI code (ACI, 1977) or the Canadian building code (CSA, 1977) using the feasible conjugate-direction method (Kirsch, 1993). The beam can be of any cross-sectional shape subjected to distributed and concentrated loads. Their cost function is similar to equation (1.1) with the exception of the term for the cost of fibers. For the examples considered they conclude that for post-tensioned members partial prestressing appears to be more economical than complete prestressing for a prestressing-to-reinforcing steel cost ratio greater than 4. For pretensioned beams, on the other hand, complete prestressing seems to be the best solution. For partially prestressed concrete they also conclude that for a prestressing-to-reinforcing steel cost ratio in the range of 0.5 to 6, the optimal solutions vary little. Cohn and MacRae (1984b) perform parametric studies on 240 simply supported, reinforced, partially, or completely pre- and post-tensioned prestressed concrete beams with different dimensions, depth-to-span ratios, and live load intensities. They conclude that, in general, RC beams are the most cost-effective at high depth-to-span ratios and low live load intensities. On the other hand, completely prestressed beams are the most cost-effective at low depth-to-span ratios and high live load intensities. For intermediate values, partial prestressing is the most cost-effective option.

Saouma and Murad (1984) present the minimum cost design of simply supported, uniformly loaded, partially prestressed, I-shaped beams with unequal flanges subjected to the constraints of the 1977 ACI code (ACI, 1977). The optimization problem is formulated in terms of nine design variables: six geometrical variables plus areas of tensile, compressive, and prestressing steel. The constrained optimization problem is transformed to an unconstrained optimization problem using the interior penalty function method (Kirsch, 1993) and is solved by the quasi-Newton method (Vanderplaats, 1984) of the IMSL library (IMSL, 1980). They found the optimum solutions for several beams with spans ranging from 6 m to 42 m, assuming both cracked and uncracked sections, and reported cost reductions in the range of 5 % to 52 %. They also conclude that allowing cracking to occur does not reduce the cost by any significant measure.

Using integer programming, Jones (1985) formulates the minimum cost design of precast, prestressed concrete simply supported box girders used in a multi-beam highway bridge and subjected to the AASHTO (1977) loading assuming that the cross-sectional geometry and the gridwork of strands are given and fixed. The design variables are the concrete strength, and the number, location, and draping of strands (moving the strands up at the end of the beam). The constraints used are release and service load stresses, ultimate moment capacity, cracking moment capacity, and release camber. The cost function includes only the first and third terms in equation (1.1).

Abendroth and Salmon (1986) present a parametric study on the sensitivity of the optimum cost of partially or fully end-restrained RC T-section beams in terms of various parameters such as allowable deflections, material strength, support conditions, and unit material costs. The constrained minimization problem is converted to an unconstrained one using an internal penalty function and solved by the quasi-Newton–Raphson method (Kirsch, 1993). The design constraints are given in the ACI code (ACI, 1983). The cost function includes the first, second, fourth, and fifth terms in equation (1.1). In addition, they add a penalty cost parameter in order to take into account the various factors associated with increased floor thickness. They assume the unit cost of stirrups (shear reinforcement) to be one and a half times the cost of longitudinal reinforcement. They found that the optimum cost is less with metallic forms than wooden forms even when the latter is used as many as four times. They state that shear reinforcement does not have a significant role in reducing the total cost and therefore may be neglected in the optimization formulation. They report a 5 % savings in the total cost by increasing the strength of concrete from 17.2 MPa (2.5 ksi) to 48.2 MPa (7.0 ksi) and 15 % savings by increasing the yield strength of steel from 275.6 MPa (40 ksi) to 516.8 MPa (75 ksi). They compare an elastic design with a partial limit state design and state that the latter does not produce significantly more economical beams.

Park and Harik (1987) present the minimum cost design of horizontally curved two-way RC slabs with rigid boundaries based on the British code (CP110, 1980) using the sequential LP method. The cost function includes the first two terms in equation (1.1). They consider the constraints on deflections, minimum effective depths, and design moments as the three dominant factors in the optimization process.

MacRae and Cohn (1987) present the optimization of prestressed concrete flat slabs based on the Canadian code for concrete structures (CSA, 1977) and the recommendation of an ACI-ASCE Committee (ACI-ASCE, 1974) using the conjugate-direction method. Notwithstanding the importance of shear in the design of flat slabs, they only consider the flexural reinforcement in the optimization formulation. They pose the problem as one of finding the reinforcing and prestressing steel in the flat slab of a given story for given concrete dimensions. Their cost function contains only the first three terms in equation (1.1). They conducted a parametric study by varying the depth-to-span ratio, live load, cable layout, and limit state and allowable tensile stresses. They conclude that using cables in clusters (groups of cable) and using high-strength steel reduce the total cost.

Prakash *et al.* (1988) present minimum cost designs of singly and doubly reinforced rectangular and T-shape RC beams, using Lagrangian and Simplex methods per limit state conditions of the prevailing Indian code. The cost function includes the first two terms in equation (1.1). They state that a two-way slab is more economical than a T beam floor for spans up to 6 m in a residential type building, whereas for heavier loads or longer spans the reverse is true.

Paul *et al.* (1990) present the minimum cost design of a modular floor system with precast prestressed voided and solid slabs simply supported on steel beams, using the general geometric programming method (Beightler and Phillips, 1976). The design is given in the British codes (BS449, 1969; CP110, 1976). The cost function includes the cost of fabrication of the slabs including the cost of concrete, prestressing steel, and forms, cost of steel beams, and the cost of erection. They conclude that for optimum cost designs the prestressing force required for a solid slab is less than that for a voided slab.

Kanagasundaram and Karihaloo (1990) describe the minimum cost design of simply supported and continuous rectangular, L- and T-section RC beams according to the Australian code (AS3600, 1988) using two different methods: sequential LP and sequential convex programming (Arora, 1989). The constraints used are stability, strength, serviceability, durability, and fire resistance. Their cost function includes the first, second, and fourth terms in equation (1.1). For the examples considered they state that the costs of concrete and the reinforcing steel are about the same but the cost of formwork is more than twice the cost of concrete and steel combined, thus concluding the significant contribution of the formwork cost to the total cost. A sensitivity analysis of the cost optimization with respect to the cost of formwork is performed and found that the minimum cost design is not affected by the variations in the relative cost of formwork.

Kanagasundaram and Karihaloo (1991a) consider the strength of concrete (f_c') as a design variable in addition to cross-sectional dimensions and the steel ratio for the cost optimization of simply supported and multi-span beams with rectangular and T-sections. The cost of concrete is related to the concrete strength through nonlinear regression analysis and using a cubic function. They conclude that higher strength concrete (up to 60 MPa) resulting in shallower sections yields more economical beams. Kanagasundaram and Karihaloo (1991b) present the minimum cost design of RC multi-span beams subjected to earth pressure, liquid pressure, wind, or earthquake loads in addition to dead and live loads. The design constraints and cost function are the same as before. The conclusions are also similar.

Ezeldin (1991) presents the minimum cost design of rectangular, reinforced fiber concrete beams with four variables: width and depth of the beam, steel fiber content and area of bending reinforcing bars. The cost function can be obtained from equation (1.2) by setting the prestressing steel (A_{pb}), compression steel (A_{sb}'), and shear reinforcement (C_{sbv}) to zero. A direct search method is used for optimization. As an extension of this work, Ezeldin and Hsu (1992) formulate the minimum cost design of rectangular, reinforced fiber concrete beams with two additional variables, cross-sectional area and spacing of stirrups, and thus the cost function includes the cost of shear reinforcement. They conclude that the variations of the costs of concrete and form appear to have a more significant influence on the minimum cost than those of steel reinforcement and fibers.

Chakrabarty (1992a) presents the minimum cost design of RC rectangular beams using the geometric programming (Kirsch, 1993) and Newton–Rapson methods. The cost function includes the first, second, and fourth terms in equation (1.1). In the context of the Indian condition where the labor is cheap, the author found that at optimum solution the cost of concrete and steel are about the same but the cost of formwork is about one-fourth of the cost of concrete or steel. The reverse was reported earlier by Kanagasundaram and Karihaloo (1990) for countries like Australia or the USA where the labor cost is high. The author observes that in most cases the optimum design yields ductile beams, which is desirable for withstanding dynamic forces like earthquakes. In an extension of this work Chakrabarty (1992b) concludes that the minimum cost design of a rectangular singly reinforced beam is increased by 36 % when the width-to-depth ratio is increased from 0.25 to 0.67. The author observes that optimized sections are often deeper sections to satisfy displacement constraints and hence become more ductile with less steel reinforcement. Erbatur *et al.* (1992) discuss the minimum cost design of prestressed concrete beams with rectangular and flanged sections. They solve the non-linear optimization problem approximately by using the LP approach.

Cohn and Lounis (1993) present the minimum cost design of partially and fully prestressed concrete continuous beams and one-way slabs. The optimization is based on the limit state design and projected Lagrangian algorithm. They simultaneously satisfy both collapse and serviceability limit state criteria based on the ACI code (ACI, 1989). The material nonlinearity is idealized by an elastoplastic constitutive relationship. A constant prestressing force and prestressing losses are assumed. Their cost function includes the first three terms in equation (1.1). They report that the total cost decreases with the increase in the allowable tensile stress (f_t).

Lounis and Cohn (1993b) present a multi-objective optimization formulation for minimizing the cost and maximizing the initial camber of post-tensioned floor slabs with serviceability and ultimate limit state constraints of the ACI code (ACI, 1989). The cost objective function is chosen as the primary objective and the camber objective function is transformed into a constraint with specified lower and upper bounds. The resulting single optimization problem is then solved by the projected Lagrangian method. The cost function for the slab includes only the first and third terms in equation (1.1).

Khaleel and Itani (1993) present the minimum cost design of simply supported partially prestressed concrete unsymmetrical I-shaped girders per ACI code (ACI, 1983). The objective function is similar to equation (1.1) but with the exception of the last term. The sequential quadratic programming method is used to solve the nonlinear optimization problem assuming both cracked and uncracked sections. They conclude that an increase in the concrete strength does not reduce the optimum cost significantly, and higher strength in prestressing steel reduces the optimum cost to a certain extent. They state that some amount of reinforcing steel facilitates the development of cracking in the concrete, which reduces the cost of materials and improves ductility.

Al-Salloum and Siddiqi (1994) present the minimum cost design of singly reinforced rectangular concrete beams per ACI code (ACI, 1989). The cost function includes only the first, second, and fourth terms in equation (1.1). They obtained a closed-form solution for the steel areas and depth in terms of the cost and strength parameters by taking the derivatives of the augmented Lagrangian function with respect to the area of steel reinforcement, depth of beam, and four Lagrange multipliers for constraints on flexural strength, lower and upper bounds on ductility, and the side constraint.

Adamu *et al.* (1994) outline a continuum-type optimality criteria approach (Rozvany *et al.* 1994) for the minimum cost design of singly reinforced RC beams with rectangular cross sections based on the European code (CEB/FIB, 1990). The cost function includes only the first, second, and fourth terms in equation (1.1). The necessary cost minimality criteria are obtained by applying the calculus of variation to an augmented Lagrangian function. They applied the method to a propped cantilever beam (fixed support at one end and simple support at the other end) with variable depth and width. As an extension of this work, Adamu and Karihaloo (1994a) used the discretized continuum-type optimality criteria (DCOC) method for the minimum cost design of RC beams with varying cross-sections using the depth or the depth and steel reinforcement ratio as design variables. They applied the method to two example problems: a propped cantilever beam and a three-span continuous beam. Adamu and Karihaloo (1994b) discuss the minimum cost design of rectangular RC beams with uniform cross-sections and variable steel ratio in each span. Adamu and Karihaloo (1995a) consider the minimum cost design of nonprismatic RC simply supported T-beams and propped cantilever rectangular beams with segmentation. In each segment of the beam, the cross-section is either constant or varies linearly or quadratically.

Han *et al.* (1995) discuss the minimum cost design of partially prestressed concrete rectangular, and T-shape beams based on the Australian code (AS3600, 1988) using the DCOC method. The cost function includes the first four terms in equation (1.1). They conclude that for a simply supported beam, a T-shape is more economical than a rectangular section. Han *et al.* (1996) use the DCOC method to minimize the cost of continuous, partially prestressed and singly reinforced T-beams with constant cross-sections within each span. A three-span and a four-span continuous beam example is presented.

1.2.2 Concrete Columns

Few papers have been published on the cost optimization of concrete columns. The general cost function for a concrete column can be written in a similar way to equation (1.1) for concrete beams:

$$C_m = C_{cc} + C_{sc} + C_{pc} + C_{fc} + C_{tc} \qquad (1.3)$$

where C_{cc}, C_{sc}, C_{pc}, C_{fc}, and C_{tc} are the costs of concrete, reinforcing steel, prestressing steel, formwork and lateral ties in columns, respectively. For pre-tensioned columns, equation (1.3) can be written as

$$C_m = w_c H_c \left(A_{cc} - A_{sc} - A_{pc} \right) c_c + w_s H_c A_{sc} c_s + w_p H_c A_{pc} c_p$$
$$+ H_c p_{fc} c_f + V_{tc} c_s \qquad (1.4)$$

where H_c is the height of the column, A_{cc} is the cross-sectional area of the column, A_{sc} is the cross-sectional area of the steel reinforcement, A_{pc} is the cross-sectional area of the prestressing steel, p_{fc} is the cross-sectional perimeter of the form, and V_{tc} is the volume of the lateral ties.

Kanagasundaram and Karihaloo (1990, 1991a) present the minimum cost design of rectangular RC columns subjected to an axial compressive force and single or biaxial bending based on the Australian code (AS3600, 1988) using sequential LP and sequential convex programming methods. Both short and long columns are considered, taking into account their slenderness ratio. The cost function is similar to equation (1.3) but without the prestressing cost. Both objective function and constraints are approximated by Taylor's series expansions. In a subsequent paper, Kanagasundaram and Karihaloo (1991b) include the concrete strength as a design variable in addition to cross-sectional dimensions and the area of the longitudinal reinforcement.

Zielinski *et al.* (1995) present the cost optimization of RC short tied rectangular columns based on the Canadian code (CSA, 1984) using the internal penalty function method. The cost function includes the first, second, and fourth terms in equation (1.3). Kocer and Arora (1996) present the minimum cost design of prestressed concrete transmission poles based on the PCI (1983) and the ACI (1977) codes using (a) a combination of branch and bound, enumeration, and sequential quadratic programming methods and (b) a genetic algorithm (Goldberg, 1989; Adeli and Hung, 1995). Their cost function includes the first, third, and fourth terms in equation (1.3). Their results indicate a genetic algorithm to be more efficient than the other approach used. They report savings in the neighborhood of 25 % compared with conventional designs.

1.2.3 Concrete Frame Structures

The overall or total cost of a concrete structure (C_T) can be expressed in the following form:

$$C_T = C_m + C_f + C_t + C_s + C_{cd} + C_e \tag{1.5}$$

where C_f, C_t, C_s, C_{cd}, and C_e are the costs of fabrication (or placement), transportation, substructure (or foundation), cladding, and erection, respectively. Only a few papers have been published on the cost optimization of reinforced concrete frame structures. All of them deal with two-dimensional frames with two exceptions.

Andam and Knapton (1980) discuss the minimum cost design of portal precast RC frames but without presenting much detail. Krishnamoorthy and Mosi (1981) present the cost optimization of two-dimensional frames with rectangular cross-sections using the sequential unconstrained minimization technique (SUMT) (Fiacco and McCormick, 1968) and Davidon-Fletcher-Powell method (Arora, 1989). They considered nonlinear constitutive relationships but no actual design code. Their cost function includes only the material costs of concrete, steel reinforcement, and formwork. They present examples of single-, double-, and triple-bay and two-, four-, and six-story frames.

Huanchun and Zheng (1985) present a two-level minimum cost design approach for two-dimensional RC frames according to the Chinese building code. In the first level they try to find the most flexible structure satisfying the global constraints, such as the lateral drift using the sequential LP method. In the second level the cost of the frame is minimized by considering the local constraints for each member of the structure and using a discrete search method for cross-sectional widths and depths. Their cost function includes the material costs for beams and columns only. Choi and Kwak (1990) minimize the costs of rectangular beams and columns of RC frames by using a direct search method to select appropriate design sections from some predetermined discrete sections based on the ACI (1977) and Korean codes. Their cost function includes the material costs of concrete, steel, and the formwork.

Spires and Arora (1990) discuss the optimal design of tall tubular RC framed structures with double symmetry in the plan based on the ACI code

(ACI, 1983) using a sequential quadratic programming procedure. However, they reduce the doubly symmetric structure into the equivalent plane frame using the approximate finite element approach proposed by Khan (1974). As such, they optimize the cost of regular symmetric two-dimensional frames. Their cost function includes the material costs of concrete, steel, and framework for beams and columns. They also consider the frequency constraint in order to limit wind and earthquake forces. They present examples of five- and forty-story two-dimensional frames.

Dinno and Mekha (1993) discuss the minimum cost design of one-and two-story RC frames based on the ACI code (ACI, 1983) using an inelastic trilinear moment–rotation relationship for beams and columns, and SUMT. They consider the material costs of concrete, reinforcement, and formwork. They conclude that optimal designs using inelastic analysis results in somewhat more economical designs.

Moharrami and Grierson (1993) present the minimum cost design of RC building frames subjected to vertical and lateral loading based on the ACI code (ACI, 1989) using the optimality criteria approach. The columns have rectangular cross-sections and the beams can be rectangular, L, or T shapes. The design variables are the width, depth, and longitudinal steel reinforcement of the beams and columns. Their cost function includes the material costs of the concrete, reinforcement, and the formwork. Their largest example is a five-story single-bay RC frame. They conclude that the optimality criteria approach converges slowly when stiffness constraints are included in the formulation.

Adamu and Karihaloo (1995b) used the discretized continuum-type optimality criteria (DCOC) method for the minimum cost design of two-dimensional multi-bay and multi-story RC frames based on Australian (AS3600, 1988) and European (CEB/FIB, 1990) limit state design codes. The cost function includes the material costs of concrete, reinforcing steel, and the formwork. The design variables are cross-sectional dimensions and steel ratios. For economical reasons they assume uniform beam and column dimensions in every story but vary the steel ratios in each member. This also reduces the cost of the formwork because the formwork can be re-used more frequently. They present the optimum cost design of a seven-story RC frame with setbacks. In a companion paper, Adamu and Karihaloo (1995c) take into account the biaxial bending of the corner columns approximately, but still considering plane frames.

Fadaee and Grierson (1996) present the minimum cost design of three-dimensional RC frames with members subjected to biaxial moments and shear forces using the optimality criteria approach based on the ACI code

(ACI, 1995). Beams and columns are assumed to have rectangular sections. The cost function includes the material costs of concrete, steel, and the formwork. The focus of this work is formulation of the appropriate constraints for combinations of the axial load, biaxial bending moment, and biaxial shear. Their example is only a one-bay and one-story space frame. They conclude that the biaxial shear is an important consideration for the design of columns and its inclusion increases the cost of the optimum structure significantly.

Balling and Yao (1997) present a comparative study of optimization of three-dimensional RC frames with rectangular columns, and rectangular, T-, or L-shape beams according to the ACI code (ACI, 1989) using one-, two-, and four-story frames subjected to vertical and lateral loads, and employing the sequential quadratic programming or a gradient-based method. For steel reinforcement they consider two different definitions for design variables. In the first definition, the area of steel in each member is the only design variable used for steel in that member. In the second definition, they consider the number, diameter, and longitudinal distribution of the reinforcing bars, and perform a two-level optimization. They include the costs of materials, fabrication, and placement in the cost function by assuming that (a) the material and fabrication costs of the steel reinforcement are proportional to the weight and (b) the placement cost is proportional to the number of bars, stirrups, and ties. They conclude that the optimum costs based on the two definitions are very close to each other and thus there is no need to include the second more computationally costly definition in the optimization formulation. Based on this conclusion, the authors then discuss a simplified approach for cost optimization of space RC frames.

1.2.4 Bridge Structures

As one of the first papers on cost optimization of structures, Torres *et al.* (1966) present the minimum cost design of prestressed concrete highway bridges subjected to AASHTO loading by using a piecewise LP method. The independent design variables are the number and depth of girders, prestressing force, and tendon eccentricity. They further define dependent design variables as the spacing of girders, tendon cross-sectional area, initial prestress, and the slab thickness and reinforcement. They claim their cost function includes the costs of transportation, erection, and bearings in addition to the material costs of concrete and steel, but do not give any detail. They present results for bridges with spans ranging from

20 ft to 110 ft (6.1 m to 33.5 m) and with widths of 25 ft (7.6 m) and 50 ft (15.2 m).

Yu *et al.* (1986) present the minimum cost design of a prestressed concrete box bridge girder used in a balanced cantilever bridge (consisting of two end cantilever and overhang spans and one middle simple span) based on the British code (CP110, 1976) and using general geometric programming (Beightler and Phillips, 1976). The cost function includes the material costs of concrete, prestressing steel, and the metal formwork. They also include the labor cost of the metal formwork, roughly as 1.5 times the cost of the material for the formwork. The design variables are the prestressing forces, the eccentricities, and the girder depths for all spans. Barr *et al.* (1989) also use the general geometric programming method to minimize the cost of a continuous three-span bridge RC slab with an overall span length of 16.6 m subjected to the constraints of AASHTO (1983) and Ohio Department of Transportation bridge design regulations (ODOT, 1982). The cost function includes the material costs of concrete and steel.

Lounis and Cohn (1993a) present the minimum cost design of short and medium span highway bridges consisting of RC slabs on precast, post-tensioned, prestressed concrete I-girders satisfying the serviceability and ultimate limit state constraints of the Ontario Highway Bridge Design Code (OHBDC, 1983). They use a three-level optimization approach. In the first level they deal with the optimization of the bridge components including dimensions of the girder cross-sections, slab thickness, amounts of reinforcing and prestressing steel, and tendon eccentricities by the projected Lagrangian method (Haftka and Gurdal, 1992). In the second level, they consider the optimization of the longitudinal layout such as the number of spans, restraint type and span length ratios, and transverse layout such as the number of girders and slab overhang length. In the third level, they consider various structural systems such as solid or voided slabs on precast I- or box girders. They use a sieve-search technique (Kirsch, 1993) for the second and third levels of optimization. Their cost function includes the material costs of concrete, reinforcement, and connections at piers. They also include the costs of fabrication, transportation, and erection of girders assuming a constant value per length of the girder. They conclude by optimizing a complete set of bridge system results in a more economical structure than optimizing the individual components of the bridge. Based on their optimization studies they recommend simply supported girders for prestressed concrete bridges of up to 27 m (89 ft) long, two-span continuous girders for span lengths of 28 m (92 ft) to 44 m (144 ft), three-span continuous girders for span lengths of 55 m (180 ft) to 100 m (328 ft), and two- or three-span continuous girders for an intermediate range of 44 m (144 ft) to 55 m (180 ft).

Cohn and Lounis (1994) apply the above three-level cost optimization approach to multi-objective optimization of partially and fully prestressed concrete highway bridges with span lengths of 10 m to 15 m and widths of 8 m to 16 m. Their objective functions include the minimum superstructure cost, minimum weight of prestressing steel, minimum volume of concrete, maximum girder spacing, minimum superstructure depth, maximum span-to-depth ratio, maximum feasible span length, and minimum superstructure camber. For a four-lane 20 m length single-span bridge, they conclude that the voided slab and the precast I-girder systems are more economical than the solid slab and one- and two-cell box girders. Lounis and Cohn (1995a) also conclude that voided slab decks are more economical than box girders for short spans (less that 20 m) and wide decks (greater than 12 m), and single-cell box girders are more economical for medium spans (more than 20 m) and narrow decks (less than 12 m). The single-cell box girder, however, results in the deepest superstructure, which may be a drawback when there is restriction on the depth of the deck. Multi-criteria cost optimization of bridge structures is further discussed by Lounis and Cohn (1995b, 1996). They suggest that the criteria of minimax and minimum Euclidean distance can be used by designers for selection of the *best* solution.

Fereig (1996) presents the minimum cost preliminary design of single-span bridge structures consisting of cast-in-place RC deck and girders based on the AASHTO code (AASHTO, 1992). The author linearizes the problem by approximating the nonlinear constraints by straight lines and solves the resulting linear problem by the Simplex method. The author concludes that 'it is always more economical to space the girder at the maximum practical spacing'.

1.2.5 Water Tanks

Saxena *et al.* (1987) present the minimum cost design of RC water tanks based on the Indian and ACI (1969) codes using the heuristic flexible tolerance method (Himmelblau, 1972). The cost function includes the material costs of concrete, steel, and the formwork. They conclude that a larger percentage in cost savings can be achieved for water tanks with larger capacities.

Using a direct search method and the SUMT, Tan *et al.* (1993) present the minimum cost design of RC cylindrical water tanks based on the British code for water tanks. The cost function includes the material costs of concrete and steel only. The tank wall thickness is idealized with piecewise linear slopes with the maximum thickness at the base.

1.2.6 Folded Plates and Shear Walls

Lakshmy and Bhavikatti (1995) present the minimum cost design of simply supported trough-type folded plate roofs based on the Indian code using a combination of sequential LP and the SUMT. The cost function includes the material costs of concrete and steel only.

Hajek and Frangopol (1991) describe a computer program for the minimum cost design of concrete shear wall systems based on the ACI code (ACI, 1983) using the folded plate theory and the method of feasible directions (Vanderplaats, 1984). The cost function includes the costs of concrete and the formwork (including transport and labor) but excluding the cost of reinforcement.

1.2.7 Concrete Pipes

Thakkar and Sridhar Rao (1974) discuss cost optimization of composite-type prestressed concrete pipes based on the Indian code. They approximate the constraints by linear functions and solve the resulting problem by the LP method. The cost function includes the material costs of concrete and steel only. Heinloo and Kaliszky (1981) present a closed-form approximate solution for the minimum material cost design of thick-walled plastically rigid RC pipes subjected to internal pressure.

1.2.8 Concrete Tensile Members

Naaman (1982) presents the minimum cost design of prestressed concrete tensile members based on the ACI code (ACI, 1977). He approximates the nonlinear optimization problem to a linear one and solves it by the LP method. The cost function includes the material costs of concrete and the prestressing steel. Optimization of a 30.48 m (100 ft) long tie member of an arch structure subjected to an axial tensile force of 444.8 kN (100 kip) is presented.

1.2.9 Cost Optimization Using the Reliability Theory

All the aforementioned references use a deterministic approach to cost optimization. A few researchers have used the reliability theory to include the uncertainties in the computation of the design loads and resistances. In deterministic optimization a structure is optimized only for a given predetermined set of loadings. In reliability-based design the loads and the structural strengths are considered as random variables, and safety is related to some probability of exceeding the structural capacity by the applied loading. In reliability-based optimization an attempt is made to consider different failure modes under different loading scenarios simultaneously. The reliability-based optimization arguably can incorporate the interactions among various failure modes. However, the major bottleneck in the reliability-based optimization is the computation of the probability of failure, which often cannot be done consistently due to insufficient statistical data.

The reliability factor in the cost optimization is considered either directly or indirectly. In the direct approach, the reliability factor is included directly in the objective function. Moses (1977) presents the total cost (C_T) as the summation of the initial cost (C_I), which is a function of design variables, and the expected failure cost (C_F) multiplied by a probability of failure (P_F), which is also considered a function of design variables:

$$C_T = C_I + P_F C_F \qquad (1.6)$$

subjected to the design constraints:

$$h_i(\mathbf{x}) = 0, \qquad i = 1, 2, \ldots, N_{ch} \qquad (1.7)$$

$$g_i(\mathbf{x}) \geq 0, \qquad i = 1, 2, \ldots, N_{cg} \qquad (1.8)$$

where N_{ch} and N_{cg} are the total number of equality and inequality constraints, respectively. The second term in equation (1.6) represents the risk of the loading on the structure exceeding its capacity. The expected failure cost includes the cost associated with the failure of the structure, such as replacement cost, damage to properties, casualties, business interruption, litigation costs, etc.

In the indirect method the objective function is only the initial cost. The reliability term is considered indirectly in the form of a constraint or constraints in addition to the design constraints, equations (1.7) and (1.8), such as

$$P_F \leq P_{F \text{ allowable}} \qquad (1.9)$$

Thus, in this approach a deterministic optimization procedure can be converted into a reliability-based optimization procedure by adding one or more additional probability constraints.

Moses (1977) uses both direct and indirect approaches for the minimum cost design of RC beams and highway girders subjected to fatigue loading using SUMT and a direct search procedure. The probability of failure is calculated from a safety index, which is in turn computed from the mean values and the standard deviations of the random strength and load parameters (Frangopol and Moses, 1994). The expected failure costs are chosen in advance somewhat arbitrarily.

Surahman and Rojiani (1983) present a reliability-based optimization of four- and ten-story RC building frames by including the reliability term in the cost function. By varying the probabilities of failures between the range 0.000 001 to 0.01 and assuming different values for the expected failure cost, they arrive at an *optimum* probability of failure. SriVidya and Ranganathan (1995) discuss the reliability-based cost optimization of single-story single-bay RC frames based on different live and wind load conditions and the Indian code. They perform an elastoplastic analysis and include both component- and system-level probabilities of failure in the form of constraints. They simply assume values for probabilities of failure. Lin and Frangopol (1996) present the reliability-based minimum cost design of simply supported RC T-girders for highway bridges based on AASHTO provisions (AASHTO, 1992). Their initial cost is only the material costs of concrete and steel. The optimization approach is the method of feasible directions. They point out that only about 4 % of the structural optimization papers are about concrete and composite structures.

Koskisto and Ellingwood (1997) present the minimum life-cycle cost optimization of prefabricated concrete structures using the reliability theory. They define the total life-cycle cost as

$$C_L = C_D + C_P + C_C + C_Q + C_M + P_F C_F \tag{1.10}$$

where C_D is the planning and design cost, C_P is the production cost, C_C is the construction cost, C_Q is the quality assurance and quality control costs, and C_M is the preventive and corrective maintenance costs. For an example problem of a hollow core slab, they assume the design cost as 2.5 % of the production cost, where the production cost is the sum of materials and the labor costs. They ignore the C_Q value and assume that the labor cost is 43 % of the material costs (C_m) and the construction cost (C_C) is 0.01 times the span length and thickness of the slab. They use the projected Lagrangian method to solve the optimization problem.

1.2.10 Concluding Comments

The great majority of papers on cost optimization of concrete structures includes the material costs of concrete, steel, and formwork. Some researchers ignore the cost of the formwork. However, this cost is significant in industrialized countries and should not be ignored. Other costs such as the cost of labor, fabrication, placement, and transportation are often ignored. Additional research needs to be done on life-cycle cost optimization of structures where the life-cycle cost of the structure over its lifetime is minimized instead of its initial cost of construction only.

The researchers of reliability-based optimization make a valid argument about the inclusion of uncertainties in loads and resistances in the optimization process. However, at present (and in the foreseeable future) the probabilities of failure and the *expected failure* costs cannot be calculated with any measure of certainty due to insufficient statistical data; they have to be chosen somewhat arbitrarily or in some magical way!

1.3 Cost Optimization of Steel Structures

In this section, a chronological review of papers is presented on the cost optimization of steel structures published in archival journals. This review is divided into three subsections: deterministic, reliability-based, and fuzzy logic-based cost optimization of steel structures. In deterministic cost optimization of steel structures, where the great majority of the papers are published, optimization is performed for a predetermined set of loadings based on code-specified constraints. In reliability-based cost optimization, loads and resistances are considered to be random and the optimization is performed for a given safety factor or probability of exceeding the structural capacity. In a fuzzy logic-based optimization an attempt is made to take into account the imprecisions in determining the cost parameters and constraints using the theory of fuzzy sets (Zadeh, 1965).

1.3.1 Deterministic Cost Optimization

For steel structures a general total cost function (C_T) can be defined in the following form:

$$C_T = C_m + C_f + C_t + C_e \tag{1.11}$$

where C_m is the material cost of structural members (beams, columns, and bracings), C_f is the fabrication cost (including the material costs of connection elements, bolts, and electrode, and the labor cost), C_t is the cost of transporting the fabricated pieces to the construction field, and C_e is the erection cost (including the material costs of connection elements, bolts, and electrode, and the labor cost).

When only the material cost of structural members is included (the first term in equation (1.11)) the cost function can be presented as proportional to the volume or weight of the structure:

$$C_m = c_m \rho_s V = c_m W \tag{1.12}$$

where ρ_s is the unit weight of steel, c_m is the cost per unit weight of steel, V is the volume of the structure, and W is the total weight of the structure. In this case the cost optimization problem is simply transformed to the weight optimization problem. This simplification also assumes that various hot-rolled shapes commonly used for beams, columns, and bracings have the same unit price, which may not be the case.

Some authors use equation (1.12) as their objective function and refer to the resulting problem as the 'cost' optimization problem. In this work, those papers are considered as a weight optimization problem and consequently excluded from this review. The review in this section is classified based on the type of steel structures.

1.3.1.1 Beams and Plate Girders

For beams and plate girders a general cost function can be defined in the following form:

$$C_T = C_{mb} + C_{fb} + C_{tb} + C_{eb} \tag{1.13}$$

where C_{mb}, C_{fb}, C_{tb}, and C_{eb} are the material, fabrication, transportation, and erection costs of the beams or plate girders, respectively. The great majority of published articles include only the first two terms in equation (1.13) in the cost optimization formulation and use the following reduced cost function:

$$C_T = C_{mb} + C_{fb} \tag{1.14}$$

An early attempt on cost optimization of steel girders is presented by Razani and Goble (1966). They optimize doubly symmetric I-shaped welded plate girders with a constant web depth based on design requirements similar to AASHO (1961) using the dynamic programming method (Adeli and Ge, 1989; Adeli, 1994). Their cost function includes both the material cost of the girder, including stiffeners and splices, and the fabrication cost. They balance the material cost with the fabrication cost to minimize the total cost by smoothening the variations in the flange thickness along the span and minimizing the use of flange splices, resulting in less welding. They present two example problems: a simple-span girder with overhangs on both ends subjected to a uniformly distributed load and a three-span continuous girder under moving AASHO (1961) loads.

Goble and DeSantis (1966) present the minimum cost design of composite, continuous welded plate girders used in highway bridges with unequal flanges and variable-thickness flange and web plates but with constant depth. The basis of design is AASHO (1961). The design variables are the flange thicknesses and widths, the web thicknesses and depth, and the distances between the web plate splices and the bottom and top flange splices. They find the minimum cost web height and flange width for a given arrangement of splice points. They mention various costs for fabrication and welding including the cost of pre-heating steel, the cost of preparing and aligning edges of the plates before welding, the cost of welding rods, and the cost of weld metal depositing (labor cost). The material cost for a steel plate is considered to be made of three components: a basic cost per pound, an extra cost per inch of thickness, and an extra cost per inch of width, assuming a higher cost for thicker and wider plates. However, the authors present no actual cost function. The optimization technique used is dynamic programming. They present an example of a two-span continuous plate girder with a span length of 60.96 m (200 ft) under moving AASHO (1961) loads. Moses and Goble (1970) point out that using similar cross-sections for many members can reduce the fabrication cost of a framed steel structure. In other words, the minimum cost structure is often somewhat heavier than the minimum weight structure. They describe the use of dynamic programming for minimum cost selection of member sizes without actually presenting any structure made of those members.

Annamalai *et al.* (1972) present the minimum cost design of simply supported welded plate girders subjected to concentrated and uniformly distributed loads based on the AISC specifications (AISC, 1969). They use commercially available plates and a discrete optimization method called 'backtrack programming' (Golomb and Baumert, 1965). They provide a table of material and labor costs for different components of the plate girder but present no cost function. For welding and splices they do not present separate material and labor costs; instead they combine both material and labor costs into one cost item. As examples, they present the cost optimization of a 36.6 m (120 ft) simply supported plate girder with and without flange splices.

Anderson and Chong (1986) present the minimum cost design of homogeneous and hybrid stiffened steel plate girders according to the AISC code (AISC, 1978). They consider two factors that raise the cost of a stiffened hybrid girder over the cost of an unstiffened homogeneous girder: (1) the additional labor cost for cutting and welding stiffeners, and (2) the cost of higher strength steel for the flange plates. They present analytical functions for the *optimum* depth of the web plate by making a number of assumptions such as neglecting the tension field effect and the shear–tension interaction.

Lorenz (1988) discusses the minimum cost design of composite beams based on the AISC Load and Resistance Factor Design (LRFD) code (AISC, 1986). He suggests that the true advantage of the LRFD code can be realized in a minimum cost design. The author is concerned with the trade-off between steel weight and the number of studs needed, without considering the cost of concrete, and presents an equivalent 'cost-rated beam weight' to take into account the costs of beam and studs for conditions limited to uniformly distributed loading, ASTM A36 steel, concrete strength of 20.7 MPa (3 ksi), and a particular size of studs.

Farkas (1991) presents closed-form solutions for optimum cost values of the cross-sectional variables for simply supported welded box girders subjected to a uniformly distributed load and simplified noncode constraints on bending stress, local flange and web buckling, shear fatigue for longitudinal fillet welds, and deflection. The cost function is similar to equation (1.14) where the cost of fabrication (C_{fb}) is expressed as a function of different labor times required for (a) preparation, assembly, and tacking, (b) welding, (c) electrode changing, weld slagging, and chipping, and (d) post-treatment of welds (toe burr grinding). The author presents empirical equations for various items using empirical data. In order to come up with the closed-form solutions, the author makes a number of simplifying assumptions, e.g. assuming the relation $b = 2\,h/3$ between the flange width (b) and web depth (h). The primary conclusion of this work is that the fabrication details and costs play an important part in the optimum cost design of welded steel structures. Further, for an example box girder with a span of 10 m subjected to a distributed load of 60 kN/m, the author reports that the minimum cost design is about 11 % more economical than the minimum weight design.

Bhatti (1996) presents the minimum cost design of simply supported partially or fully composite I-shaped steel beams with concrete slabs subjected to a uniformly distributed load, and strength, deflection, and vibration constraints of the AISC LRFD specifications (AISC, 1994) using the Lagrange multiplier approach. The resulting equations are solved by using a symbolic algebra program such as Mathematica (Wolfram, 1988). The cost function is similar to equation (1.14). However, the fabrication cost includes the cost of field-installed studs only. The cost optimization is formulated in terms of the relative cost of field-fabricating a stud to the cost per pound of the rolled steel (a ratio varying in the range of 6 to 12). Graphical solutions of several examples with span lengths of 7.6 m and 12.2 m are presented.

1.3.1.2 Trusses

Lipson and Russell (1971) discuss the minimum cost design of a roof structural system consisting of welded parallel-chord trusses, purlins, deck, and wall cladding above the bottom chord based on the Canadian code (CSA, 1965). The top and bottom chord members are T-section, and web members are double angles. The design variables are member sizes, spacing of trusses, depth-to-span ratio of trusses, number of panels of trusses, and the spacing of purlins. Their cost function includes the cost of materials for trusses, decking, purlins, and the wall cladding, and the cost of fabrication including the costs of preparation of chord and web members, splicing, welding, and labor. The labor cost is expressed in terms of the number of members rather than the weight of the truss. The cost of wall cladding is expressed as a step function of truss depth and spacing. The costs of decking and purlins are expressed as step functions of the spacing of trusses and purlins. The optimization approach is a modified Simplex method for a nonlinearly constrained optimization problem dubbed the 'Complex' method (Box, 1965). Lipson and Gwin (1977) discuss the minimum cost design of steel space trusses subjected to the AISC constraints (AISC, 1970). Their cost function includes the first two terms in equation (1.11). The fabrication cost includes the cost of galvanization. The optimization approach is the same 'Complex' method. They present a 25-member space truss example made of steel angles.

Thomas and Brown (1977) discuss the cost optimization of a truss roof system consisting of a number of identical one-way two-dimensional trusses, open-web joists, and standard 22 gage decking materials subjected to the AISC specification (AISC, 1970). Their cost function includes the first, second, and fourth terms in equation (1.11) for the aforementioned components. The material and erection costs of the roof decking are assumed to be proportional to the roof area. The fabrication and erection costs of both the open-web joists and primary trusses are assumed to be proportional to their weights. The optimization method is the sequential unconstrained minimization technique (SUMT) (Arora, 1989) and the Davidon–Fletcher–Powell method (Fletcher and Powell, 1963). The largest example truss presented has 37 members and covers a span of 32.6 m (1283.5 in).

Imai (1983) presents a mini-max dual approach for minimum weight and cost optimization of trusses made of steel and aluminum members subjected to explicit displacement and stress constraints. The theoretical idea is to combine the lightweight but more expensive aluminum with heavier but less expensive steel in an economical way. The cost function is the sum of the scaled material costs and weights of the components of a structure. The author acknowledges the difficulty in dealing with the discontinuous nature of the material properties and approximates the problem using the first-order Taylor series expansion for a displacement response. A 72-bar space truss example is presented.

1.3.1.3 Plane Frames

Ridha and Wright (1967) discuss the minimum cost design of two-dimensional steel frames using the mechanism method of the simple plastic analysis and the plastic design requirements of the AISC (1963) code assuming adequate bracing against buckling in the weak axis direction. The cost function includes the first two terms in equation (1.11) but only the cost of welded connections is included in the fabrication cost. The authors assume that the connection cost is a linear function of the shear force and the bending moment resisted by the connection. Further, the connection cost includes another component as a function of the size of the connected members intended to represent the costs of detail drawing, making templates, shear angles, and reaming. They report that compared with the minimum weight design the minimum material and connection cost design results in a heavier frame but a lower total cost, indicating the relative importance of the connection cost. A two-bay and two-story frame and a single-bay and three-story frame are presented as examples. They report savings in the range of 7 % to 26 % for the minimum cost design versus the minimum weight design.

Anderson and Islam (1979) attempt to present approximate closed-form solutions for the minimum cost design of multi-story rectangular rigid frames with limiting values on the lateral deflections. They oversimplify the problem by a number of assumptions, including neglecting the effect of vertical loads on lateral displacements and assuming inflection points at the midpoints of beams and columns. As such the frame becomes statically determinate and only one tier is considered for optimization.

Crawford and Jenkins (1980) present the minimum cost design of seven different types of steel single-span gable frame roof structures based on the British code (BSI, 1977) using a combination of the Complex method mentioned earlier and the pattern search of Hooke and Jeeves (1961). The single-span gable structures consist of two steel columns and a roof made of hot-rolled steel sections with or without haunches, plate girders, Warren truss, or trussed beam. Only the roof structure is optimized, excluding the columns, purlins, and sheeting. The authors studied the relative cost advantages of the various roof structures for a span range of 10 m to 50 m in the construction environment of the United Kingdom and provide relative cost curves and recommendations for practicing engineers. They also present curves for optimum length-to-span ratios and the number of panels versus the span length fitting through data. This paper demonstrates how cost optimization algorithms can directly help practicing engineers.

Majid *et al.* (1980) present the minimum cost topological design of rigid frames subjected primarily to lateral deflection constraints. The nonlinear optimization problem is approximated linearly by the Taylor series expansion and solved by the Simplex method. The cost function is the summation of the material cost and a constant value representing roughly the construction cost. They present examples of two-, three-, and five-story and multi-bay frames. Topological optimization is carried by simply removing some of the columns. Nakamura and Takenaka (1983) also discuss an analytical method for the minimum cost design of rectangular multi-story multi-span frames without considering any actual code constraints. Their cost function includes the first two terms in equation (1.11), but the fabrication cost includes the cost of connections only. Such highly limited analytical solutions have academic values only.

Douty (1980) describes the minimum cost design of three different types of bolted and welded connections used in steel frames based on the AISC specifications (AISC, 1970): shear angle-framed connection, and flange and end plate moment-resisting connections. The cost function is presented in terms of the connection variables, such as the diameter of the bolts, flange plate width and thickness, shear plate length and thickness, and the leg size of the fillet weld for connecting the shear plate to the column flange and for the flange plate moment connection. A weighting penalty is included in the size of the welds and bolt diameter, assuming that the cost is increased for larger size bolts and welds. The nonlinear programming problem is approximately linearized using the Taylor series expansion and then solved by a linear programming approach. Cheng and Juang (1989) present the minimum cost design of multi-story rigid frames subjected to static wind and earthquake forces according to the Uniform Building Code (UBC, 1984). They include the PΔ effect in the formulation and solve the problem using the optimality criteria approach (Adeli, 1994). They present empirical functions for costs of members, painting, and welded connections. Their examples include a two-bay, fifteen-story rigid frame.

Thurston and Sun (1993) present the multi-criteria optimization of two-dimensional steel frames without using any actual design code constraints. They attempt to minimize both cost and lateral drift using a combination of the Pareto optimization approach (Koski, 1994) and the multi-attribute utility theory. The cost function is presented as a function of the length of the steel members and volume of the concrete used in a rectangular floor deck. An example of a one-bay, three-story frame is presented.

Xu and Grierson (1993) present the minimum cost design of steel frames with semi-rigid connections based on the AISC code (AISC, 1978) using the augmented Lagrangian method. The cost function includes the material cost of the members (proportional to their weights) and the cost of each connection, assumed to be proportional to its rotational stiffness. Examples of one-bay and two-story, and three-bay and ten-story steel frames are given. Based on the limited examples and the aforementioned assumption about the cost of connections, they report that for low-rise frames with insignificant lateral displacements, semi-rigid connections 'may sometimes' result in a lighter design compared with the more common rigid connection design. However, the total cost of the semi-rigid frame may be more than that of the corresponding rigid frame because the authors did not include the actual fabrication cost of semi-rigid connections (including the labor cost). When lateral loads dominate the design, such as in the case of tall frames, the authors state that the fully rigid design will probably yield a lighter design because it provides a greater lateral stiffness. Examples of optimal cost designs of semi-rigid, low-rise industrial frames are also given in Xu *et al.* (1995).

Simoes (1996) presents the minimum cost design of semi-rigid steel frames subjected to stress and displacement constraints but without using an actual design code. The nonlinear programming problem is approximated by the Taylor series expansion and solved by the segmented linear programming approach. The cost of the members is assumed to be proportional to the weight. The cost of connections is taken as a quadratic function of the connection fixity factor in the range of 0 (for simple pinned connections) to 1 (for moment-resisting connections) and empirical ad hoc values are used for the coefficients of the cost function. The pinned and moment connections are assumed to add 20 % and 60 %, respectively, to the cost of each member, and the additional cost of semi-rigid connections is assumed to fall within this range. The largest example presented is a two-bay, three-story semi-rigid frame. For the small low-rise frames presented and for the assumed cost functions the authors assert that both the weight and cost of a semi-rigid frame are less than those of the corresponding moment-resisting frame.

1.3.1.4 Industrial Buildings

Bradley *et al.* (1974) discuss the minimum cost design of one-story industrial framed structures using the simple plastic theory and geometric programming technique (Beightler and Phillips, 1976; Abuyounes and Adeli, 1986) without using any actual design code. Their focus is computation of the cost terms. Lee and Knapton (1974) also describe their investigation of the minimum cost design of industrial building structures made of steel portal frames based on the British code (BSI, 1969) using the simple plastic theory but without presenting an explicit cost function. The design variables are the number of bays, frame spacing, eaves height, roof pitch, purlin spacing, and building length and width. The simplified optimization problem is solved approximately by a revised Simplex method.

Russell and Choudhary (1980) present the minimum cost design of one-story industrial buildings made of roof trusses in the transverse direction, braced frames in the longitudinal direction, and the footings under the columns based on the Canadian code (CSA, 1975). The problem is first decomposed into three optimization subproblems. Then, three interface variables are defined as the number of panels in the transverse trusses, the number of bays in the longitudinal direction, and the depth-to-span ratio of trusses, and the overall cost optimization problem is solved by using the aforementioned Complex method (Box, 1965). The cost function includes the costs of materials, labor, equipment, overheads, and profit. Profit and overheads are included as a fraction of the other three direct costs. The costs of labor and equipment are presented as functions of man-hours needed in various operations. Empirical equations are presented for times required for various operations as functions of design parameters such as the number of braced frames, the number of panels in the transverse trusses, and the truss weight, based on the curve fitting of the previous data in the Canadian construction environment. The authors present optimization of an industrial building covering a rectangular $30.5\,m \times 159\,m\,(100\,ft \times 520\,ft)$ area and the clear height to the underside of the trusses of $7.6\,m$ (25 ft).

Jendo and Paczkowski (1993) describe the single- and multi-objective minimum cost design of one-story industrial buildings made of roof space double-layer trusses consisting of tubular sections subjected to explicit constraints on displacements, stresses, and buckling, using the metric and utility function methods (Jendo, 1990). Similar to Russell and Choudhary (1980), the problem is decomposed into several subproblems for optimization of roof covering (purlins and corrugated sheet), space trusses, columns, and walls (corrugated sheets). They attempt to synthesize the various optimization subproblems by two global or interface variables: the height of the trusses and the 'mesh density', defined as the ratio of the span length to the distance between the truss nodes. The multi-criteria are the minimization of the weight of the truss structure and the wall and column elements, the maximum vertical displacement, and the labor cost expressed empirically.

1.3.1.5 Guyed Towers

Bell and Brown (1976) discuss a heuristic approach for the minimum cost design of cable-supported steel guyed towers with a height in the range 30 m to 150 m used for supporting heavy microwave antennas subjected to wind loading and the AISC specifications (AISC, 1970) by assuming independence of design variables in various subspaces of the design. Optimum cable areas and tower mast sections are found independently using the Powell search method (Powell, 1964) and the branch and bound algorithm. They consider only the material costs and for simplicity transform the minimum cost design problem to a quasi-minimum weight design problem by assuming a fixed ratio for the relative costs of the cable and tower steel. The design variables are cable cross-sectional area, initial cable tension, mast cross-sectional area, and anchor and tie locations. The cross-section of the mast is either square with four angles or triangular with three angles. They do not seem to include the weight of the cross bracings in the formulation. However, the nonlinear behavior of the cables is included in the formulation.

1.3.1.6 Steel Transmission Poles

Kocer and Arora (1997) formulate the cost optimization of self-supporting steel transmission poles made of two overlapping tapering dodecagonal tubes with constant thickness. In addition to the dead load of the pole, the National

Electric Safety Code's light loading, ASCE ice and wind loading, and broken conductor loading are considered. The constraints in the design are given in the ASCE guidelines (ASCE, 1990). The design variables are the outside diameter at the top of the pole, tapering of the pole, and thicknesses of the two overlapping pieces. The cost function is similar to equation (1.14) with the fabrication cost formulated as the welding cost with three different cost items. They are the costs of total labor and overheads, total electrode used, and the power and equipments needed for welding. They calculate the total labor and overhead costs for welding from the length of the weld, hourly labor and overhead charge, welding done by one worker per hour, and a so-called operating factor. The power and equipment cost is assumed to be 20 % of the total electrode costs. The authors include the secondary moment effects due to lateral displacements in the formulation and solve the problem using three different approaches: (a) genetic algorithm (Adeli and Cheng, 1993, 1994a; Adeli and Hung, 1995), (b) simulated annealing (Aarts and Korst, 1989), and (c) the enumeration method. By including the additional labor costs, the optimum values of the diameter at the top of the upper tube and the thicknesses of both tubes are increased, and the optimum value of the tapering slope is decreased. They report that the genetic algorithm is the best of the three methods used in terms of computational efficiency and finding the global optimum solution.

1.3.1.7 Cellular Plates

Farkas and Jarmai (1994) present the minimum cost design of laterally loaded welded cellular steel plates using three different approaches: the backtracking method, the hill-climbing method, and feasible sequential quadratic programming (Farkas and Jarmai, 1997). The cellular plate is created by sandwiching and welding a grid of cold-formed channels or I-beams between two parallel plates. The design constraints are defined explicitly for bending stresses and local buckling of rib webs due to bending and shear without using any actual design code. The cost function includes the material and fabrication costs. The latter is calculated by multiplying the total time required for fabrication by a fabrication cost factor. The total time required for welding is the sum of the times required for (a) preparation, assembly, and tacking, (b) welding, and (c) electrode changing, weld deslagging, and chipping. Empirical equations based on local fabrication conditions are used for various conditions. They conclude that the hill-climbing approach is quick but sensitive to initial solutions and the feasible sequential quadratic programming is 'robust' even when the starting point is infeasible.

1.3.1.8 Bridge Structures

Memari *et al.* (1991) present the minimum cost design of a continuous, multispan concrete reinforced concrete–steel girder highway bridge subjected to the AASHTO code (AASHTO, 1983) using the method of feasible directions. The cost function is expressed as the material costs of the superstructure including the costs of the steel girders, longitudinal and transverse stiffeners, studs, and the reinforced concrete slabs. Their unit costs of materials are intended to include other costs such as fabrication, transportation, and erection indirectly. An example of a three-span and two-lane bridge structure is presented. They conclude that the dimensions of the flange and web plates have the greatest impact on the minimum cost solution.

1.3.2 Cost Optimization Using the Reliability Theory

Papers published on the reliability-based cost optimization of structures all take an academic, theoretical, and idealistic approach to the problem. The examples presented in these publications are usually small, academic two-dimensional structures. None of them uses an actual widely used design code such as the AISC specifications (AISC, 1995). Use of the probabilistic concepts in structural design was presented by Benjamin (1968). One of the first papers published on the reliability-based structural cost optimization is Mau and Sexsmith (1972). They minimize the expected cost of simple statically determinate two-dimensional steel trusses as defined by equation (1.6). They make a number of simplifying assumptions such as limiting each member to only one type of failure and ignoring partial failure and serviceability criterion. The initial cost of the structure, the first term in equation (1.6), is taken as the material cost only, which is expressed as a function of the weight of the structure. The cost of failure is assumed known and taken as proportional to the initial material cost of the structure. They point out that the criterion of minimum expected cost is equivalent to minimization of weight with an allowable probability of failure.

Ravindra and Lind (1973) describe the use of the probability theory in design code optimization with an attempt to balance the safety and cost. They apply the concept by finding a set of optimal load factors for single-story single-bay steel frames subjected to dead, snow, and wind loads using the hill climbing approach (Rosenbrock, 1960). Moses (1977) introduces the general concepts of the reliability theory into structural optimization in the context of the two approaches presented in Section 1.2.9. Rao (1980) presents the minimum cost design of a cable-stayed cantilevered steel box beam with a probabilistic objective function and constraints using the indirect approach. The external loadings and the ultimate stresses are considered as random variables. The author transforms the stochastic formulation to an equivalent deterministic nonlinear programming problem by assuming that the random variables follow a normal distribution with small standard deviations, expanding the objective function about the mean values of the random variables by Taylor's series expansion, and approximating the series by the first two terms. The equivalent deterministic nonlinear programming problem is solved by SUMT (Arora, 1989) with an interior penalty function. A probability of failure of 0.0001 is assumed.

Frangopol (1985) gives two reasons why the reliability-based structural optimization has not been popular as compared with the deterministic structural optimization. First, the lack of a universally acceptable method for incorporating the uncertainties in the structural optimization formulation results in nonuniform reliability levels in similar structural design situations. Second, the diverging opinions on many basic issues include the very definition of reliability-based optimization. The author then advocates a multi-criteria optimization approach with collapse and unserviceability as the failure criteria. The method is applied to a single-story rigid steel frame with random strengths and random vertical and horizontal concentrated loads, assuming 0.000 01 and 0.01 for probabilities of collapse and unserviceability, respectively.

Soltani and Corotis (1988) present single- and multi-objective formulations with initial and failure costs as objectives functions. Design variables are the mean plastic moment capacities of structural members using the simple plastic theory for steel structures. They define the cost of failure as the replacement cost and the cost of compensation for possible damage caused by failure and note that the evaluation of this cost is extremely difficult, especially if human lives are endangered. The multi-objective optimization problem is solved by the so-called 'constraint method', where one of the objectives is treated as a constraint with lower and upper bound limits (Cohon, 1978). The approach is applied to a one-story single-bay steel rigid frame with initial cost treated as an additional constraint.

Kim and Wen (1990) present the reliability-based cost optimization of structures under multiple stochastic (time-varying) loads. The combined effects of the loads, treated as random processes, are included using the load coincidence method (Pearce and Wen, 1984). The optimization problem is solved using SUMT (Arora, 1989) with an interior penalty function. Examples of one-story, single-bay and two-story, two-bay steel frames are presented. Enevoldsen and Sorensen (1994) present the reliability-based cost optimization of structures with component and system reliability constraints, and apply the concepts to a very simple example, a simply supported tubular steel column. Chang *et al.* (1994) discuss reliability-based cost optimization of steel structures subjected to seismic loading of the Uniform Building Code (UBC, 1988) and Newmark's nondeterministic seismic response spectra (Paz, 1991) and apply it to a ten-story, single-bay steel frame assuming rigid floors. The optimization problem is solved by SUMT (Arora, 1989). They conclude that nonstructural costs as well as future failure costs can affect structural cost only at high failure probability levels. Tao *et al.* (1995) use the Markov decision process and structural reliability theory to model the minimum expected lifetime cost of a structure and apply the concepts to a composite five-girder highway bridge.

1.3.3 Fuzzy Optimization

Fuzzy optimization is based on the theory of fuzzy sets developed by Zadeh (1965). Brown and Yao (1983) introduce the application of the fuzzy set theory in structural engineering and state:

> It has been argued that probability theory and statistics are useful in civil engineering but their use is limited in the sense that most civil engineering decisions are made with a shortage of numerical evidence and depend on informed opinions. The fuzzy set theory is intended to deal with the informed opinion, but in no way disperses with countable evidence.

Reliability-based optimization is based on the long-established theory of probability while fuzzy optimization is based on the more recent theory of possibility (Zadeh, 1978) based on the theory of fuzzy sets (Zadeh, 1965). Probability is based on the premise that events or variables are random in nature with a statistical basis, but possibility is based on a fuzzy domain with mostly nonstatistical variables. In the fuzzy optimization, numerical values of the membership functions are used, as opposed to the probabilities in the reliability-based optimization. In structural design, two major sources of fuzziness, imprecision, or uncertainties can be identified, one in the evaluation of the structural behavior and resistance, the other in determining the loadings acting on the structure. In the cost optimization of structures, a third source of fuzziness and imprecision comes into play. That is in the formulation and evaluation of the cost function.

A fuzzy set Y for any set Z is characterized by a membership function $\mu_Y(z)$ which grades each point in Z with a value in the interval $[0,1]$. This membership function is the grade of membership of z in Y. The nearer the value of $\mu_Y(z)$ to unity the higher is the grade of membership of z in Y. Thus, a fuzzy set Y is defined as

$$Y = \{z, \mu_Y(z)\} | z \in Z \tag{1.15}$$

The fuzzy set theory can be used to model judgments on ambiguous, imprecise, or fuzzy situations. The membership functions of a fuzzy set are used to develop a fuzzy transition from total acceptance to total rejection of certain decision processes.

A number of papers have been published on fuzzy optimization of structures (Wang and Wang, 1985a, 1985b; Rao 1987a, 1987b; Yeh and Hsu, 1990; Rao *et al.*, 1992a, 1992b, 1992c; Yu and Xu, 1994; Shih and Lai, 1994). Most of these papers, however, are on weight optimization. Only a few deal with the cost optimization of structures presenting academic examples. In these papers, the cost function for the fuzzy cost optimization of structures is expressed as the summation of the initial cost (C_I) and the expected cost of maintenance and failure (C_E):

$$C_T = C_I + C_E \tag{1.16}$$

This equation is somewhat similar to equation (1.6) for reliability-based optimization.

Wang and Wang (1985a) present a simplified fuzzy optimization procedure, dubbed the α-level cut method, by considering the fuzziness in the constraints and using the nonfuzzy cost function defined by equation (1.16). The membership functions for the constraints are restricted to preselected lower limit values of α. As such, the amount of fuzziness in the constraints is limited to preselected ranges. The advantage of this approach is that the problem is readily transformed to ordinary nonfuzzy optimization with expanded lower and upper bound limits, which are functions of α. The disadvantage of this approach is that the α values are selected somewhat arbitrarily. The authors apply the concepts to two small academic examples, a one-bay, two-story shear frame and a three-bar truss. Wang and Wang (1985b) discuss a two-step approach for fuzzy optimum design of aseismic structures considering both construction cost and earthquake-caused loss expectation during the service life of the structure. They introduce the concept of a fuzzy response spectrum and apply the approach to a simple one-bay, two-story shear frame. Yeh and Hsu (1990) also discuss a similar procedure for the cost optimization of structures with fuzzy allowable strength and fuzzy loads using the simple plastic theory and theory of possibility (Zadeh, 1978). They assume exponential functions for the fuzzy allowable strength and fuzzy loads, and apply the concepts to a simple three-bar truss and a one-story, one-bay frame.

1.3.4 Concluding Comments

Only a small fraction of structural steel optimization articles attempt to include any cost other than the weight of the structure. Most of the cost optimization papers are applied to small or academic examples. With the exception of a few that present moderate-size problems, all are really small-scale optimization problems.

A number of articles have been published on the reliability-based cost optimization of steel structures. Practically all present simple academic examples. Nearly three decades ago Moses (1977) acknowledged the weak database for determining statistical parameters needed in a meaningful reliability-based cost optimization. The same problem of an inadequate database exists even today and will exist in the foreseeable future. Another problem is the existence of a large number of possible failure modes, especially for large structures, which makes the evaluation of system reliability in a consistent practical way an impossible task. While the reliability theories can make a real contribution in advancing the development of more realistic design codes, their use in practical cost optimization of realistic structures appears limited at the present time.

So far only a few articles have been published on the cost optimization of steel structures using the fuzzy set theory, which deal with small academic examples. The authors of these papers appear to have been influenced by the reliability-based optimization in formulating the problem. The approach is primarily the α-level cut method, which includes the fuzziness in the constraints only. However, there are significant sources of fuzziness in the cost function as well, and the fuzzy set approach provides an effective way of modeling them.

It is interesting to note that there was a relatively good amount of research activity in the cost optimization of structures in the 1960s and 1970s. This activity dwindled in the 1980s and picked up again in the 1990s to some extent. The cost optimization problem is somewhat ill-defined in a mathematical sense, and in general its solution is less amenable to established algorithmic procedures and is computationally more intensive. With widespread availability of increasingly powerful personal computers and workstations and the development of recent computational paradigms such as the theory of fuzzy sets, structural optimization researchers need to pay closer attention to the cost optimization problem.

Research on cost optimization can encourage the use of the optimization approach in the structural steel design practice for at least two reasons. First, it provides a more realistic way of modeling structural steel design. Second, the consensus of the existing literature is that cost optimization can result in additional savings in the order of 7% to 26% compared to the weight optimization problem. These savings can be very significant for large structures.

For the structural optimization methodology in general, and the cost optimization approach in particular, to be embraced by the structural engineering community, the focus of research should be on large structures subjected to the actual constraints of a commonly used design code such as the AISC ASD (AISC, 1995) or the AISC LRFD (AISC, 2001) codes. The true benefit of optimization is realized for large structures with hundreds of members.

An optimization algorithm that works for a small problem, or a large problem but with simplified constraints, may not work for a large structure subjected to the actual highly nonlinear, implicit, and discontinuous constraints of an actual design code such as the AISC LRFD code (AISC, 2001). This significant issue is hardly discussed in the structural optimization literature. It should be known that nonlinear optimization algorithms are highly sensitive to the nature of the constraints and the size of the problem. An algorithm that works for explicit simplified constraints can produce unstable results for complicated implicit and discontinuous constraints. Recently, however, new promising algorithms have been created for solution of large-scale and complicated optimization problems that produce stable results consistently, such as the recently patented neural dynamics model of Adeli and Park (Adeli and Park, 1996; Park and Adeli, 1997a, 1997b; Adeli and Park, 1998) and the evolutionary computing and genetic algorithm that will be presented in subsequent chapters.

2

Evolutionary Computing and the Genetic Algorithm*

2.1 Overview and Basic Operations

Many mathematical linear and nonlinear programming methods have been developed for solving optimization problems during the last three decades. However, no single method has been found to be entirely efficient and robust for all different kinds of engineering optimization problems. Some methods, such as the penalty function method, the augmented the Lagrangian method, and the conjugate gradient method, search for a local optimum by moving in a direction related to the local gradient. Other methods apply the first- and second-order necessary conditions to seek a local minimum by solving a set of nonlinear equations. For the optimum design of large structures, these methods become inefficient due to a large amount of gradient calculations and finite element analyses.

These methods usually seek a solution in the neighborhood of the starting point similar to local hill climbing. If there is more than one local optimum in the problem, the result will depend on the choice of the starting point, and the global optimum cannot be guaranteed. Furthermore, when the objective

---

* This chapter is based mostly on the following articles of the senior author: H. Adeli and N. T. Cheng, Integrated genetic algorithm for optimization of space structures, *Journal of Aerospace Engineering*, ASCE, 1993, **6**(4), 315–328, and H. Adeli and N. T. Cheng, Augmented Lagrangian genetic algorithm for structural optimization, *Journal of Aerospace Engineering*, ASCE, 1994, **7**(1), 104–118; and is reproduced by permission of the publisher.

---

function and constraints have multiple or sharp peaks, the gradient search becomes difficult and unstable.

Inspired by Darwin's theory of evolution and the natural law of the survival of the fittest, the genetic algorithm (GA) is a global search procedure for gradually improving the solution in succeeding populations using operations that mimic those of the natural evolution such as reproduction, crossover, and mutation (Holland, 1975; Goldberg and Samtani, 1986; Goldberg, 1989) and performs a random information exchange to create superior offsprings. Adeli and Cheng (1993) presented the optimization of space structures by integrating GA with the penalty function method. Subsequently, Adeli and Cheng (1994a) presented an improved augmented Lagrangian GA for the optimization of space structures, where the problem of the trial-and-error selection of the initial value for the penalty function coefficient is avoided. This chapter presents a summary of that work.

A simple GA has the following characteristics:

- All the genetic algorithm operations work with finite-length binary strings (chromosomes) instead of real parameter sets, resulting in a finite point-search algorithm.
- Genetic algorithms consider a group of points in the search space in every iteration, called a population of points.
- Genetic algorithms use a random search based on the prior information to guide the search, instead of gradient search, so that the derivative information and step-size calculation are not necessary.
- Genetic algorithms must work in a bounded space for coding the parameters.
- Genetic algorithms are not hill-climbing algorithms. The so-called local hill-climbing problems are eliminated in these algorithms. Therefore, the probability of being entrapped in a local minimum is reduced.

Three characteristics of a GA are attractive for optimization of large space structures. First, it starts from a group of points (design) instead of one single starting point, and guarantees a fast convergence to a near optimum solution. Second, it requires only simple function evaluations, resulting in a numerically efficient algorithm. Finally, it readily lends itself to parallel processing (Adeli, 1992a) because each chromosome or string in the population is independent of the others and therefore can be processed concurrently.

The genetic algorithm can be used directly only for solving unconstrained optimization problems. Therefore, a constrained optimization problem must be transformed to an unconstrained optimization problem, e.g. by employing the penalty function method. It will subsequently be shown how the penalty function method can be combined with the GA to solve the structural optimization problem. There are two fundamental steps in a genetic algorithm. The first step is coding and decoding of the parameter sets by bit strings. The second step is the search algorithm in order to find the optimum solution.

2.2 Coding and Decoding

Coding the parameter sets as a finite-length string over some finite alphabets is the first step in the genetic algorithm. Finite-length binary strings are usually used for decoding. The length of binary strings depends on the required accuracy. For example, a real variable X whose range is $0.0 \leq X \leq 1000.0$ can be coded as a 10-digit string:

$$0000000000 \leq X \leq 1111111111 \tag{2.1}$$

There are a total of $2^{10} = 1024$ points in this range. As an example, the point 0000011100 represents

$$C = 0.0 + \frac{2^4 + 2^3 + 2^2}{2^{10}} \times (1000.0 - 0.0) = 27.3438$$

based on the following formula for coding and decoding:

$$C = C_{\min} + \frac{B}{2^L}(C_{\max} - C_{\min}) \tag{2.2}$$

where $C =$ value the string represents, $C_{\min}$ and $C_{\max} =$ lower and upper bounds for C, $L =$ length of the binary string, and $B =$ decimal integer value of the binary string.

Structural optimization problems are always multi-parameter problems. To code the multiple parameters, the parameters can simply be concatenated end by end and then presented as a single string. For example, if there are four parameters $X_1, X_2, X_3,$ and X_4 and 10 bits are used to code a single parameter, the string will contain 40 bits. For example, for four randomly selected $X_1 = 1100011010$, $X_2 = 0010011101$, $X_3 = 1010001101$, and $X_4 = 0111101110$, the string becomes

$\underline{1100011010}$	$\underline{0010011101}$	$\underline{1010001101}$	$\underline{0111101110}$
X_1	X_2	X_3	X_4

Genetic algorithms work with the discrete points coded by the finite-length binary strings and not by the parameters themselves. Hence, they are not dependent on the continuity of the parameter space. This feature makes GAs more flexible and efficient than conventional search techniques.

2.3 Basic Operations in Genetic Algorithms

Although many genetic algorithms with different strategies have been reported, all of them consist of three basic operations: (1) reproduction, (2) crossover, and (3) mutation. Reproduction is simply a process to decide which strings should survive and how many copies of them should be produced in the mating pool. The decision is made by comparing the fitness of each string with the average fitness of the population. The fitness is an indicator of the survival potential and reproduction capability of the string in the subsequent generations. For an optimization problem, the fitness is the objective function or a combination of the objective function and constraints. In maximization problems, a string with a greater fitness will receive correspondingly more copies in the new population. On the other hand, in minimization problems, a string with a smaller fitness receives more copies in the mating pool. Suppose the population size to be n, the fitness of the ith individual string in the current iteration to be f_i, the summation of the fitnesses in the current iteration to be f_{sumi}, and the average fitness of the current iteration to be $f_{\mathrm{ave}} = f_{\mathrm{sumi}}/n$. Then, the probability of the ith individual string to be selected into the mating pool is

$$ps_i = \frac{f_i}{f_{\mathrm{sumi}}} \tag{2.3}$$

and the number of copies that the ith individual string receives is determined by

$$\mathrm{num}_i = n \times ps_i = \frac{f_i}{f_{\mathrm{ave}}} \tag{2.4}$$

These are the so-called survival-of-the-fittest aspects of the GA. The better strings (those with a smaller value of the fitness in the minimization problems) receive more copies for mating so that their desirable characters and string patterns or schemata (Holland, 1975) may be passed on to their offsprings.

Consider the following simple maximization problem:

$$\text{Maximize } F(X) = X^2 + 2X \text{ subject to } 0.0 \le X \le 63.0 \tag{2.5}$$

Suppose a binary string with a length of six is used to code the real variable X and the population size is set to be four. Using a random process, four starting points 011111, 111000, 001000, and 100001 are chosen. The four strings represent the real values 31.0, 56.0, 8.0, and 33.0, respectively. Their fitnesses, values of the function F, are 102.0, 3248.0, 80.0, and 1155.0, respectively, and the corresponding numbers of copies these strings receive are, theoretically, 0.74, 2.36, 0.06, and 0.84. Thus, by rounding off these numbers, there will be one copy of 011111, two copies of 111000, one copy of 100001, and no copy of 001000 in the mating pool. In practice, reproduction is done at random. A range is created according to the fitness of each individual. Thus, a better string will occupy a bigger portion in the range and consequently has more chance to be chosen into the mating pool.

Crossover is a means for two high-fitness strings (parents) to produce two offsprings by mixing and matching their desirable qualities through a random process. After reproduction, the crossover proceeds in two steps. First, two newly reproduced strings are selected at random for mating from the mating pool. Next, some bits chosen at random are exchanged between these two strings. Several methods can be used for choosing portions to be swapped. In this section, one-point, two-point, and uniform crossover are described.

To perform one- and two-point crossovers, one and two crossing sites along the string are selected at random, as follows: for one-point crossover

011|111

111|000

and for two-point crossover

0|111|11

1|110|00

In the foregoing examples, there are five possible crossing sites. Then, if one segment is swapped between the two strings, two offsprings are produced as follows: for one-point crossover

011000

111111

and for two-point crossover

011011

111100

The uniform crossover is based on a randomly created binary string, called a mask (Syswerda, 1989). A mask acts like a sieve. Parent strings are asked to exchange the bits on the positions where the corresponding position in the mask is zero. Otherwise, no exchange of bits is performed. The percentage of exchanged bits between two parent strings can be varied from 0 % to 50 % by controlling the percentage of zeros in the mask string. The following example shows how the 40 % uniform crossover operation works (note that the mask string contains 40 % zeros):

parent string

1000110101 0011101100

mask

1100110110

offspring

1011111100 0000100101

Although the crossover is done by random selection, it is not the same as a random search through the search space. Since it is based on the reproduction process just described, it is an effective means of exchanging information and combining portions of high-fitness solutions.

Reproduction and crossover are very simple operations. Their implementation simply requires generating random strings, making copies of the strings, and swapping portions of the strings. However, reproduction and crossover together give genetic algorithms much of their power.

The third operation in genetic algorithms, called mutation, plays an important role as a safeguard. Mutation occurs with a small probability in the genetic algorithm to reflect the small rate of mutation existing in the real world. Some digits at a particular position in all strings may be eliminated during the reproduction and crossover operations. It is impossible to recover from such a situation by using only reproduction and crossover operations. For example, in the following eight-string population, the fifth position (from the left) contains a 0 in all strings and consequently character 1 would never be produced on the position five if only reproduction and crossover operations are used:

| 110001 | 101001 | 000100 | 111101 |
| 010101 | 000001 | 111000 | 011100 |

This situation may create an additional constraint in the search domain and may prevent the desirable solution from being obtained. To avoid such a deadlock situation, in the mutation phase some bits will be changed in all the strings according to the mutation rate.

2.4 GA with the Penalty Function Method

2.4.1 Problem Formulation for Axial Force (Truss) Structures

The structural optimization of a space truss structure is defined as selecting the vector of the cross-sectional areas $\mathbf{A} = \{A_1, \ldots, A_N\}$ that minimizes the total weight W of the structure

$$W = \sum_i^N \rho_i L_i A_i = \mathbf{L}^T \rho \mathbf{A} \qquad (2.6)$$

subject to the following stress, displacement, and fabricational constraints:

$$\sigma^L \leq \sigma \leq \sigma^U \qquad (2.7)$$

$$\delta^L \leq \delta \leq \delta^U \qquad (2.8)$$

$$\mathbf{A}^L \leq \mathbf{A} \leq \mathbf{A}^U \qquad (2.9)$$

where $\rho = $ a diagonal matrix of weight per unit volume, $N = $ number of design variables, $L_i = $ length of element i, and $\mathbf{A}^U$, $\mathbf{A}^L$, σ^U, σ^L, δ^U, and $\delta^L = $ upper and lower bounds of the vectors of cross-sectional areas $\mathbf{A}$, element stresses σ, and nodal displacements δ. The nodal displacement and element stresses are found by a finite element structural analysis for every string in every iteration using the following equations:

$$\mathbf{K}\delta = \mathbf{P} \qquad (2.10)$$

$$\sigma = \mathbf{T}\delta \qquad (2.11)$$

where $\mathbf{K} = $ a structure stiffness matrix, $\mathbf{P} = $ a nodal force vector, and $\mathbf{T} = $ a transformation matrix relating nodal displacements to element stresses.

Since GAs can be used directly only for solving unconstrained optimization problems, the constrained problems are transformed to an unconstrained problem by including a penalty function (Arora, 1989). In this section a quadratic penalty function is used, and the corresponding unconstrained optimization problem becomes

$$\min \phi(\mathbf{A}) = \frac{1}{L_f} \sum_i^N \rho_i L_i A_i + \alpha \left\{ \sum_i^N \left[\left(\frac{|\sigma_i|}{|\sigma_i^a|} - 1 \right)^+ \right]^2 \right.$$
$$\left. + \sum_i^M \left[\left(\frac{|\delta_i|}{|\delta_i^a|} - 1 \right)^+ \right]^2 \right\} \tag{2.12}$$

where the last term is the penalty function, $\alpha = $ a penalty coefficient, $L_f = $ a factor for normalizing the objective function, $M = $ number of degrees of freedom, $\sigma_i = $ stress in member i, $\delta_i = $ displacement in the direction of degree of freedom i, and

$$\left(\frac{|\sigma_i|}{|\sigma_i^a|} - 1 \right)^+ = \max \left(\frac{|\sigma_i|}{|\sigma_i^a|} - 1, 0 \right) \tag{2.13}$$

$$\left(\frac{|\delta_i|}{|\delta_i^a|} - 1 \right)^+ = \max \left(\frac{|\delta_i|}{|\delta_i^a|} - 1, 0 \right) \tag{2.14}$$

$$\sigma_i^a = \sigma_i^L \qquad \text{when } \sigma_i < 0 \tag{2.15a}$$

$$\sigma_i^a = \sigma_i^U \qquad \text{when } \sigma_i \geq 0 \tag{2.15b}$$

$$\delta_i^a = \delta_i^L \qquad \text{when } \delta_i < 0 \tag{2.16a}$$

$$\delta_i^a = \delta_i^U \qquad \text{when } \delta_i \geq 0 \tag{2.16b}$$

The normalizing factor L_f is necessary here to make the terms in the objective and penalty functions dimensionally consistent. Also, the magnitude of this factor is chosen such that the two terms on the right-hand side of equation (2.12) become numerically close to each other, so that one term does not dominate the other. In the GA terminology, equation (2.12) is the fitness function, which is used in the reproduction phase.

2.4.2 Genetic Algorithm with the Penalty Function Method

The optimization process via the genetic algorithm can be divided into the following steps:

Step 1. Create a chromosome or string (design) population for the first iteration randomly (note that the size of the string population should be even).

Step 2. Select the penalty function coefficient α and the normalizing factor L_f.

Step 3. Decode each chromosome or string (design) and perform finite element structural analysis to find the element stresses and nodal displacements for each string (design).

Step 4. Calculate the fitness of each string using equation (2.12), which combines the total weight of the structure with the penalty function. Because this is a minimization problem, in this step the fitness is rescaled using the following formula:

$$\phi(A) = D_{\max} - \phi(A) \qquad \text{when} \phi(A) < D_{\max} \qquad (2.17a)$$

$$\phi(A) = 0 \qquad \text{when} \phi(A) \geq D_{\max} \qquad (2.17b)$$

The parameter $D_{\max}$ is the upper limit of $\phi(A)$. $D_{\max}$ is set equal to the average fitness for the population so that the strings with fitnesses greater than or equal to this value are discarded with no chance of their entering the mating pool. Thus, the lower fitness point receives a higher probability of survival.

Step 5. Reproduce strings (designs) into the mating pool according to the fitness calculated in step 4. In this step, each rescaled fitness corresponding to a string (design) is divided by the summation of the rescaled fitness and consequently scaled to a value between 0.0 % and 100.0 %.

Step 6. Randomly match the strings (designs) in the mating pool, two at a time, and apply crossover and mutation operations to create new offsprings (new designs). If the uniform crossover is used, a mask should be created randomly at the beginning of each iteration.

Step 7. Replace old strings by the offsprings and go to step 2 until the stopping criterion is met.

The following values are recommended by De Jong (1975) for the probability of crossover (P_c) and mutation (P_m):

$$P_c = 0.8 \tag{2.18a}$$

$$P_m = 0.005 \tag{2.18b}$$

In a GA with a penalty function method, the selection of the penalty function coefficient is very critical from the convergence point of view. Numerical experiments indicate that when a small value is used for this coefficient the solution usually converges to an infeasible solution. This is because the contribution of the penalty function in minimizing the fitness function will be small. On the other hand, when a large value is chosen for the penalty function coefficient the solution will oscillate. This can be explained by the fact that minimization of the fitness function would largely reflect the minimization of the penalty function at the cost of neglecting the objective function.

To improve the robustness of the genetic algorithm, a variable penalty function-coefficient strategy can be used; i.e. the value of this coefficient is increased after every certain number of iterations. What value to use for the coefficient and at which iteration to boost the value requires experience with the optimization problem domain. Numerical experimentation can provide an insight for judicious selection of these parameters.

A genetic algorithm can be used as an effective tool for many engineering optimization problems. Since gradient information is not required the method does not suffer from the hill-climbing problem encountered in the gradient-based mathematical programming algorithms. This problem is the entrapment of the solution near a local optimum in the vicinity of the starting point. In other words, the solution obtained from a nonlinear mathematical programming algorithm in general depends on the starting point and is only a local optimum.

In the genetic algorithm approach to engineering optimization, a near-optimum solution is found within a given accuracy, depending on the chosen length of the string (chromosomes). Furthermore, a genetic algorithm often provides an efficient approach to a moderate size engineering optimization problem because it requires only simple function evaluations. A near-optimum solution can be found after a limited number of iterations.

One problem arises when integrating a GA with the penalty function method; i.e. the solution usually goes into the infeasible region right after the first, or first few, iterations. This is a characteristic of the penalty function methods. In subsequent iterations, the solution will approach the feasible region from the infeasible region. There is no guarantee that the solutions found in a given iteration are feasible, and the optimum solution can still vary with the iteration number. This is due to the given accuracy (string length) and the potential of the random process in the genetic algorithm. Therefore, the constraints must be examined when a solution is selected for design among the population in that iteration. In practice, this check does not have to be made in every iteration, but only in the final iteration or the last few iterations.

2.5 Augmented Lagrangian Method

The augmented Lagrangian (Lagrange multiplier) method has been shown to be superior to the ordinary penalty function method (Powell, 1969; Fletcher, 1975; Belegundu and Arora, 1984). In contrast to manual control of the penalty function coefficient, an outer loop is used in the augmented Lagrangian method to adjust the coefficients automatically at some specific points using the information about constraints.

An optimum design problem with N design variables, N_{ch} inequality, and N_{cg} equality constraints can be defined as:

$$\text{Minimize } f(\mathbf{A}) \text{ subject to } g_i(\mathbf{A}) \leq 0, \quad i = 1, 2, \ldots, N_{ch} \quad (2.19)$$

$$h_i(\mathbf{A}) = 0, \quad i = 1, 2, \ldots, N_{cg} \quad (2.20)$$

where $\mathbf{A} = (A_1, A_2, \ldots, A_n)$ = the vector of design variables. For axial force (truss) structures $\mathbf{A}$ is the vector of cross-sectional areas.

Powell (1969) combined the penalty function method with the primal-dual method and reported a new penalty function in the following form for finding a local minimum of the equality constrained problem:

$$P[h(\mathbf{A}), \boldsymbol{\gamma}, \boldsymbol{\omega}] = \frac{1}{2} \sum_{i=1}^{N_{ch}} \gamma_i [h_i(\mathbf{A}) + \omega_i]^2 \quad (2.21)$$

where $\gamma_i > 0$ and ω_i = real parameters associated with the ith equality constraint. The constrained problem is converted into an unconstrained problem by minimizing the following function:

$$\varphi(A, \boldsymbol{\gamma}, \boldsymbol{\omega}) = f(A) + P[h(A), \boldsymbol{\gamma}, \boldsymbol{\omega}] \quad (2.22)$$

This is the so-called Lagrange multiplier or the augmented Lagrangian method and $\gamma_i \omega_i = \Gamma_i$ is called the Lagrange multiplier of the constraint $h_i(\mathbf{A}) = 0$.

Fletcher (1975) modified Powell's Lagrange multiplier method for solving the inequality constrained optimization problem by introducing the penalty function:

$$P\left[g(\mathbf{A}), \boldsymbol{\gamma}, \boldsymbol{\omega}\right] = \frac{1}{2} \sum_{i=1}^{N_{ch}} \gamma_i \left\{ [g_i(\mathbf{A}) + \omega_i]^+ \right\}^2 \qquad (2.23)$$

In this case, the constrained problem is converted into an unconstrained problem by minimizing the following function:

$$\varphi\left(\mathbf{A}, \boldsymbol{\gamma}, \boldsymbol{\omega}\right) = f(\mathbf{A}) + \frac{1}{2} \sum_{i=1}^{N_{ch}} \gamma_i \left\{ [g_i(\mathbf{A}) + \omega_i]^+ \right\}^2 \qquad (2.24)$$

where

$$[g_i(\mathbf{A}) + \omega_i]^+ = \max\left[0, g_i(\mathbf{A}) + \omega_i\right] \qquad (2.25)$$

The well-known quadratic penalty function method can be derived from equations (2.24) and (2.25) by setting $\omega_i = 0 (i = 1, \ldots, N_{ch})$ and the optimization process is completed by increasing the penalty function coefficients $\gamma_i (i = 1, \ldots, N_{ch})$ to infinity. However, large values of penalty function coefficients cause ill-conditioning in the solution of the optimization problem and result in slow convergence or numerical instability. Powell (1969), on the other hand, introduced an outer loop to update the Lagrange multipliers $\Gamma_i = \gamma_i \omega_i$ automatically according to the information obtained in previous iterations. There is no need for penalty function coefficients or Lagrange multipliers to go to infinity in order to ensure convergence. Furthermore, derivatives of the objective function and constraints are not necessary in the coefficient updating process.

2.6 GA with the Augmented Lagrangian Method

2.6.1 Problem Formulation for Axial Force (Truss) Structures

In an augmented Lagrangian GA method, the penalty function for optimization of a structure from equations (2.6) through (2.9) can be defined as

$$P\left(\boldsymbol{\sigma}, \boldsymbol{\delta}, \boldsymbol{\gamma}, \boldsymbol{\omega}\right) = \frac{1}{2} \left\{ \sum_i^N \gamma_i \left[\left(\frac{|\sigma_i|}{|\sigma_i^a|} - 1 + \omega_i \right)^+ \right]^2 \right.$$

$$\left. + \sum_i^M \gamma_{i+N} \left[\left(\frac{|\delta_i|}{|\delta_i^a|} - 1 + \omega_{i+N} \right)^+ \right]^2 \right\} \qquad (2.26)$$

As earlier, N = number of design variables and M = number of degrees of freedom. In this case the corresponding unconstrained optimization problem becomes

$$\text{Minimize } \phi\left(\mathbf{A}, \boldsymbol{\gamma}, \boldsymbol{\omega}\right) = \frac{1}{L_f} \sum_i^N \rho_i L_i A_i + \frac{1}{2} \left\{ \sum_i^N \gamma_i \left[\left(\frac{|\sigma_i|}{|\sigma_i^a|} - 1 + \omega_i \right)^+ \right]^2 \right.$$

$$\left. + \sum_i^M \gamma_{i+N} \left[\left(\frac{|\delta_i|}{|\delta_i^a|} - 1 + \omega_{i+N} \right)^+ \right]^2 \right\} \qquad (2.27)$$

where L_f = a factor for normalizing the objective function, σ_i = the stress in member i, δ_i = the displacement in the direction of the degree of freedom i, and the limiting values of stresses and displacements are as equations (2.15a) through (2.16b). Furthermore,

$$\left(\frac{|\sigma_i|}{|\sigma_i^a|} - 1 + \omega_i \right)^+ = \max \left(\frac{|\sigma_i|}{|\sigma_i^a|} - 1 + \omega_i, 0 \right) \qquad (2.28)$$

$$\left(\frac{|\delta_i|}{|\delta_i^a|} - 1 + \omega_{i+N} \right)^+ = \max \left(\frac{|\delta_i|}{|\delta_i^a|} - 1 + \omega_{i+N}, 0 \right) \qquad (2.29)$$

In the GA terminology, equation (2.27) is called the fitness function, which is used in the reproduction phase in order to guide the genetic search.

2.6.2 Genetic Algorithm with the Augmented Lagrangian Method

A hybrid structural optimization algorithm is presented by integrating the genetic algorithm with the augmented Lagrangian method in a nested loop. The outer loop is used to update the Lagrange multipliers according to the augmented Lagrangian method, similar to the first algorithm of Belegundu and Arora (1984). The inner loop performs the genetic algorithm to minimize the penalized objective function associated with the Lagrange multipliers in the outer loop. The hybrid algorithm is presented in seven steps. The inner loop is represented by step 3.

Step 1. Set $I = 1$ and $K = \propto$, initialize the values of vectors $\boldsymbol{\gamma}$ and $\boldsymbol{\omega}$, and choose the values of parameters $\alpha > 1$, $\beta > 1$, and $\varepsilon > 0$ and the normalizing factor L_f, where $I = $ a counter for parameter set $\boldsymbol{\omega}$ and $\varepsilon = $ the stopping criterion for the outer loop (desired accuracy).

Step 2. Generate the chromosome or string (design) population, $A_j^{(1)}(j = 1, \ldots, n)$, for the first iteration randomly, where $n = $ an even number representing the size of the string population.

Step 3. Set the counter of the inner loop, J, equal to zero and perform the genetic search to minimize $\varphi(A, \boldsymbol{\gamma}, \boldsymbol{\omega}^{[I]})$ subject to the lower and upper bounds of A.

  (a) Set $J = J + 1$.

  (b) Decode each chromosome or string (design) using equation (2.2) and find the element stresses and nodal displacements for each string using finite element structural analysis.

  (c) Calculate the fitness of each string using equation (2.27), which combines the objective function with the penalty function. Since this is a minimization problem, rescale the fitness of each string using the following formulas:

  $$\varphi(A, \boldsymbol{\gamma}, \boldsymbol{\omega}^{[I]}) = D_{\max} - \varphi(A, \boldsymbol{\gamma}, \boldsymbol{\omega}^{[I]})$$
  $$\text{when } \phi(A, \boldsymbol{\gamma}, \boldsymbol{\omega}^{[I]}) < D_{\max} \qquad (2.30a)$$

  $$\varphi(A, \boldsymbol{\gamma}, \boldsymbol{\omega}^{[I]}) = 0 \qquad \text{when } \phi(A, \boldsymbol{\gamma}, \boldsymbol{\omega}^{[I]}) \geq D_{\max} \quad (2.30b)$$

  so that the strings with fitnesses greater than or equal to the value of $D_{\max}$ are discarded with no chance of entering the mating pool. Thus, the smaller fitness string receives a higher probability of survival. In this work, $D_{\max}$ is set equal to the average fitness for the population.

  (d) Reproduce strings (designs) into the mating pool according to the rescaled fitness just calculated. Each rescaled fitness corresponding to a string is divided by the summation of the rescaled fitnesses and consequently scaled to a value between 0.0 and 100.0 %. Thus, better strings occupy bigger portions on the range and consequently receive more copies during the reproduction phase. Then, n numbers between 0.0 % and 100.0 % are chosen randomly and compared with the aforementioned range in order to select n preferred strings (designs) and include them in the mating pool.

(e) Match the strings (designs) in the mating pool randomly, two at a time, and apply crossover and mutation operations to create new offsprings (new designs). If the uniform crossover is used, as discussed earlier, a mask should be created randomly at the beginning of each iteration.

(f) Replace old strings by the offsprings and go to the first part of step 3 until the stopping criterion (for the inner loop) is met or $J = J_{\max}$. The population of new strings is represented by $\mathbf{A}^{(I)} = (\mathbf{A}_1^{(I)}, \ldots, \mathbf{A}_n^{(I)})$, and the string (design) with the smallest fitness in this population is represented by $\mathbf{A}^{(I)*}$.

Step 4. Evaluate the values of constraints, $g_i\left(A_j^{(I)}\right)$ $(i = 1, \ldots, N + M)$ for every string and calculate the average value for each constraint i as follows:

$$g_{i_{\text{ave}}}^{(I)} = \frac{\sum\limits_{j=1}^{n} g_i\left[A_j^{(I)}\right]}{n} \tag{2.31}$$

Set

$$K^* = \max_i \left|\max\left[g_{i_{\text{ave}}}^{(I)}, -\omega_i^{(I)}\right]\right| \tag{2.32}$$

and

$$I = \left\{i : \left|\max\left[g_{i_{\text{ave}}}^{(I)}, -\omega_i^{(I)}\right]\right| > \frac{K}{\alpha}\right\} \tag{2.33}$$

If $K^* \leq \varepsilon$ (stopping criterion for the outer loop), then terminate the run and $\mathbf{A}^{(I)*}$ is the solution. Otherwise go to the next step. Note that there is no guarantee that the solution found here is feasible, and the constraints must be examined when the solution is selected for design among the population.

Step 5. If $K^* \geq K$, update γ_i and $\omega_i^{(I)}$ using the following equations for all $i \in \mathbf{I}$, increase the counter I by one, and go to step 3:

$$\gamma_i = \beta\gamma_i \tag{2.34}$$

$$\omega_i^{(I)} = \frac{\omega_i^{(I)}}{\beta} \tag{2.35}$$

Otherwise go to the next step.

Step 6. Set

$$\omega_i^{(I+1)} = \omega_i^{(I)} + \max\left[g_{i_{\text{ave}}}^{(I)}, -\omega_i^{(I)}\right] \quad (i = 1, \ldots, N+M) \quad (2.36)$$

If $K^* \leq K/\alpha$, set $K = K^*$ and $I = I + 1$, go to step 3; otherwise go to the next step.

Step 7. Update $\gamma_i = \beta\gamma_i$ and $\omega_i^{(I+1)} = \omega_i^{(I+1)}/\beta$ for each $i \in \mathbf{I}$, and $K = K^*$. Then, increase the counter I by 1 and go to step 3.

An attractive characteristic of GA is that there is no line search and the problem of computation of derivatives of the objective function and the constraints is avoided. This feature of GA is maintained in the augmented Lagrangian GA. Compared with the penalty function-based GA, only a few additional simple function evaluations are needed in the augmented Lagrangian GA. Furthermore, instead of a single penalty function coefficient a set of Lagrange multipliers is used in this algorithm. In a sense each constraint is assigned its own penalty function coefficient, and the coefficients are adjusted individually. Thus, the imbalance of penalties from some specific constraints are avoided, resulting in a rapid and stable convergence history.

In the augmented Lagrangian GA, the wild-guess or trial-and-error approach for the starting penalty function coefficient and the process of arbitary adjustments are avoided. There is no need to perform an extensive numerical experimentation for finding a suitable value for the penalty function coefficient for each type or class of optimization problem. Thus, the augmented Lagrangian GA can be applied to a broad class of optimization problems. For examples of applications of this algorithms refer to Adeli and Cheng (1994a, 1994b).

3

Cost Optimization of Composite Floors*

3.1 Introduction

Composite floor construction is widely used in commercial multi-story buildings. An economical design is often achieved by attaching a concrete deck to the top (compression) flange of steel beams to carry the maximum positive moments. To create a composite floor, a concrete slab is often mechanically connected to a hot-rolled steel section by shear studs.

In practice a composite beam is usually designed by trial-and-error selection of the following parameters: concrete type expressed by its compressive strength (f'_c) and unit weight (γ_c), slab thickness (t_c), beam spacing (b), slab steel reinforcement (A'_s), steel section size expressed by its cross-sectional area (A_s), steel grade expressed by its yield stress (F_y), the type of shear stud expressed by its shear strength (Q_n), and the number of shear studs (N_s).

The design of composite beams is complicated and highly iterative. Depending on the design parameters a beam can be fully composite or partially composite. In the case of design on the basis of the Load and Resistance Factor Design (LRFD) code (AISC, 2001) one has to consider the plastic deformations. A source of complexity is due to the fact that the location of

---

* This chapter is based on the following articles of the senior author: H. Adeli and H. Kim, Cost optimization of composite floors using neural dynamics model, *Communications in Numerical Methods in Engineering*, 2001, **17**, 771–787, and H. Kim and H. Adeli, Discrete cost optimization of composite floors using a floating-point genetic algorithm, *Engineering Optimization*, 2001, **33(4)**, 485–501; and is reproduced by permission of the publishers.

---

the plastic neutral axis (PNA) can be within the concrete slab, the flange of the steel beam, or the web of the steel beam. Since the value of a design parameter affects other values, all design parameters cannot be found simultaneously. In practice, slab thickness and the number of shear studs are often chosen by engineers somewhat arbitrarily.

Mathematical optimization provides a methodology to automate the complicated design process (Adeli, 1994). Further, one can achieve an optimum solution out of numerous solutions on the basis of a selected criterion such as minimum weight or minimum cost. Zahn (1987) discusses the economies of the LRFD code versus the AISC Allowable Stress Design (ASD) code (AISC, 1995) in the design of composite floor beams through weight comparison of some 2500 composite designs using A36 steel. He concludes that for short-span beams in the range of 10 ft (3.05 m) and 20 ft (6.1 m) the vibration serviceability constraint is the controlling design constraint using either one of the codes. The author's 'preliminary results' indicate that the LRFD code yields a saving of '6 % to 15 % for span lengths ranging from 10 to 45 ft (3.05 to 13.7 m)'.

In this chapter, a general formulation for the cost optimization of composite beams is presented by including the costs of (a) concrete, (b) steel beam, and (c) shear studs. The resulting optimization problem is then solved by two approaches: (1) the genetic algorithm and (2) the neural dynamics model of Adeli and Park (Adeli and Park, 1995a, 1995b, 1998; Park and Adeli, 1997a, 1997b).

3.2 Minimum Cost Design of Composite Beams

3.2.1 Cost Function

A total cost function is defined in the following form:

$$C_T = C_c + C_s + C_{sd} + C_o \tag{3.1}$$

where C_c, C_s, C_{sd}, and C_o are the costs of concrete, steel beam, shear studs, and other related work, respectively. The first three terms are defined by

$$C_c = Lbt_c C_c' \tag{3.2}$$

$$C_s = \rho A_s L C_s' \tag{3.3}$$

$$C_{sd} = N_s C_{sd}' \tag{3.4}$$

where L = the beam span length, ρ = the unit weight of steel, C_c' = the cost of concrete per unit volume, including the placement cost, C_s' = the cost of the steel beam material and installation per unit weight, and C'_{sd} = the cost of installing one shear stud including the material cost.

Cost of other related work includes the costs of forms, finishing floor, and curing with sprayed membrane curing compound and others. This related cost, however, is mostly a function of the floor area (Means, 1999), which is given and fixed at the beginning in each design and, consequently, can be dropped from the optimization formulation. Further, for the purpose of optimization the unit cost coefficients C'_c, C'_s, and C'_{sd} can be replaced with equivalent cost rates normalized to one of the cost coefficients, say C'_s. Thus, a total cost function can be defined as

$$C_T = Lbt_c \frac{C'_c}{C'_s} + \rho A_s L + N_s \frac{C'_{sd}}{C'_s} \tag{3.5}$$

or, for simplicity,

$$C_T = Lbt_c C_{sc} + \rho A_s L + N_s C_{ss} \tag{3.6}$$

where C_{sc} = the relative cost of the unit volume of concrete to the cost of the unit weight of steel and C_{ss} = the relative cost of a shear stud to the cost of the unit weight of steel. This cost function is an expansion of the function defined by Lorenz (1988) with an additional term for the cost of concrete. In equations (3.5) and (3.6), the design variables are t_c, A_s, and N_s.

3.2.2 Constraints

All the design constraints are included from the AISC LRFD specifications described briefly in the following paragraphs.

3.2.2.1 Flexural Strength Constraints

The ultimate bending moment must be less than or equal to the nominal flexural strength times the resistance factor. Two cases must be considered. First, the ultimate bending moment capacity of the noncomposite steel section (excluding the concrete strength) must be checked to make sure that the steel beam can support the weight of the wet concrete, temporary loads such as construction loads, and its own weight. This constraint is expressed as

$$M_{u\_noncomposite} \leq 0.90 M_{n\_noncomposite} \tag{3.7}$$

where $M_{u\_noncomposite}$ = the required ultimate moment capacity for the non-composite steel section and $M_{n\_noncomposite}$ = the nominal moment capacity of the noncomposite steel section.

Second, the ultimate bending moment capacity of the composite section to carry all the required dead and live loads must be checked, as defined by the following constraint:

$$M_{u\_composite} \le 0.85 M_{n\_composite} \tag{3.8}$$

where $M_{u\_composite}$ = the required ultimate moment capacity for the composite beam and $M_{n\_composite}$ = the nominal moment capacity of the composite beam. The nominal bending strength in equation (3.8) is computed from a plastic stress distribution. The case of partially composite beams is considered, while the case of fully composite beams can be deduced as a special case of the partial composite beam.

Using the notation of Figure 3.1, the nominal bending stress of the composite beam when PNA is in the beam flange is expressed as

$$M_{n\_composite} = C_{con}\left(x_1 + x_2 + t_c - \frac{a}{2}\right) + C_{flange}\left(\frac{x_1}{2} + x_2\right) \tag{3.9}$$

where a = the depth of the equivalent rectangular stress block of concrete, given by

$$a = \frac{C_{con}}{0.85 f_c' b_{eff}} \tag{3.10}$$

Since the concrete compression capacity of a partially composite beam is governed by shear studs, the compression force in concrete, C_{con}, is substituted by the shear strength carried by shear studs between points of maximum and zero bending moments. Thus, equation (3.10) can be rewritten as

$$a = \frac{\sum Q_n}{0.85 f_c' b_{eff}} = \frac{Q_n N_s / 2}{0.85 f_c' b_{eff}} \tag{3.11}$$

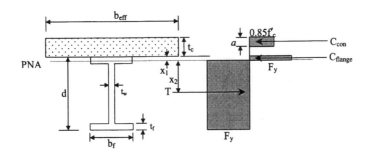

Figure 3.1   Plastic stress distribution for the PNA in the beam flange

where Q_n = the strength of a single shear stud and N_s = the total number of shear studs.

The distance between the bottom of the concrete slab and the PNA, x_1, can be found from the equilibrium between the tension force and total compression forces:

$$x_1 = \frac{A_s F_y - \sum Q_n}{2 F_y b_f} \tag{3.12}$$

The distance from the PNA to the centroid of area in tension, x_2, is obtained as

$$x_2 = \frac{A_s(d/2 - x_1) + b_f x_1^2/2}{A_s - b_f x_1} \tag{3.13}$$

When the PNA is in the beam web, as shown in Figure 3.2, the nominal bending stress of the composite beam is

$$M_{n\_composite} = C_{con}(x_1 + x_2 + t_c - a/2) + C_{flange}(x_1 + x_2 - t_f/2)$$
$$+ C_{web}[x_2 + (x_1 - t_f)/2] \tag{3.14}$$

Similarly, x_1 and x_2 are calculated by

$$x_1 = t_f + \frac{A_s F_y - \sum Q_n - 2C_{flange}}{2 F_y t_w} \tag{3.15}$$

$$x_2 = \frac{A_s(d/2 - x_1) + b_f t_f(x_1 - t_f/2) + (x_1 - t_f)^2 t_w/2}{A_s - b_f t_f - t_w x_1 + t_f t_w} \tag{3.16}$$

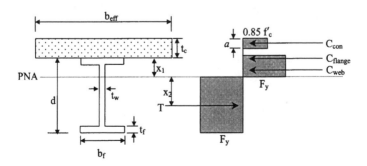

Figure 3.2 Stress distribution for the PNA in the beam web

3.2.2.2 Deflection Constraints

The AISC LRFD code does not include any explicit requirement on deflections. However, the deflection constraint is included for the generality of the optimization formulation. The deflection of a composite beam depends on whether it is shored during the construction. Shoring provides a temporary support during the hardening of the concrete slab, and consequently reduces the deflection of the composite beam. However, the unshored construction is less labor intensive and faster than the shored construction, and is consequently often the preferred method of construction. For unshored composite beams, the deflection of the composite beam due to live load, Δ_{LL}, is limited to a certain value defined as a percentage of the span length in the following form:

$$\Delta_{LL} = \frac{5w_{LL}L^4}{384E_s I_{tr}} \leq c_1 L \tag{3.17}$$

where w_{LL} = the uniform service live load weight per unit length of beam, E_s = modulus of elasticity for steel, I_{tr} = moment of inertia of the transformed fully composite section, and c_1 = a numeric coefficient such as 1/360.

3.2.2.3 Floor Vibration Constraint

The AISC LRFD code defines the floor vibration constraint in terms of an allowable damping ratio in the following form:

$$35A_0 f + 2.5 \leq D \tag{3.18}$$

where D = the percentage of critical damping, A_0 = the maximum initial amplitude of the floor system due to a heel-drop excitation in inches, and f = the first natural frequency of the floor system in Hz. Heel-drop excitation measurements proposed by Murray (1991) identify the natural frequencies and damping ratios of each floor. The response of each floor for a single person walking across the floor at critical frequencies generating resonance is measured and analyzed.

In equation (3.18), the first natural frequency of a composite simply supported beam is a function of the total weight of the beam, W, the transformed moment of inertia, and the beam span length as follows:

$$f = 1.57 \left(\frac{gE_s I_{tr}}{WL^3} \right)^{0.5} \tag{3.19}$$

where $g =$ the gravitational acceleration. The maximum initial amplitude of the floor system, A_0, is then obtained by the following equation:

$$A_0 = \begin{cases} A_1 & \text{when } t_0 \geq 0.05 \\ A_2 & \text{when } t_0 < 0.05 \end{cases} \tag{3.20}$$

where

$$A_1 = \frac{1}{N_{\text{eff}}} \left(0.246 L^3 \frac{0.10 - t_0}{E_s I_{\text{tr}}} \right)$$

$$A_2 = \frac{1}{N_{\text{eff}}} \left\{ \frac{0.246 L^3}{E_s I_{\text{tr}}} \left[\frac{1}{2\pi f} \sqrt{2(1 - a \sin a - \cos a) + a^2} \right] \right\}$$

$$t_0 = \frac{1}{\pi f} \tan^{-1} a \qquad \text{and} \qquad a = 0.1/(\pi f)$$

The number of effective beams, N_{eff}, is computed from

$$N_{\text{eff}} = 2.967 - 0.05776 \frac{b}{d_e} + 2.556 \times 10^{-8} \frac{L^4}{I_{\text{tr}}} + 0.0001 \left(\frac{L}{b} \right)^3 \geq 1.0 \tag{3.21}$$

where $d_e =$ the effective depth of the slab. The parameter d_e is equal to the average slab thickness when the metal deck ribs are perpendicular to the beam and equal to the concrete thickness above the metal deck when the deck ribs are parallel to the beam.

3.2.2.4 Maximum/Minimum Shear Stud Spacing Constraint

The AISC LRFD code defines the minimum center-to-center spacing of shear studs as not to be less than six times the diameter, and the maximum center-to-center spacing as not to be greater than eight times the total slab thickness:

$$\text{stud\_spacing} \geq 6 \times \text{stud\_diameter} \tag{3.22}$$

$$\text{stud\_spacing} \leq 8 \times t_c \tag{3.23}$$

The typical use of shear studs is shown in Figure 3.3.

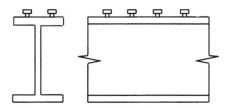

Figure 3.3   Shear stud

3.2.2.5 Constraint Normalization

Constraints are normalized to narrow down the constraint variations and consequently avoid numerical ill-conditioning and improve the optimization performance. This also prevents numerically large constraints from dominating the computation and causing entrapment of the solution in the vicinity of a local minimum:

$$\frac{M_{u\_noncomposite}}{0.90M_{n\_noncomposite}} - 1 \leq 0 \tag{3.24}$$

$$\frac{M_{u\_composite}}{0.85M_{n\_composite}} - 1 \leq 0 \tag{3.25}$$

$$\frac{\Delta_{LL}}{c_1 L} - 1 \leq 0 \tag{3.26}$$

$$\frac{35A_0 f + 2.5}{D} - 1 \leq 0 \tag{3.27}$$

$$1 - \frac{stud\_spacing}{6 \times stud\_diamer} \leq 0 \tag{3.28}$$

$$\frac{stud\_spacing}{8 \times t_c} - 1 \leq 0 \tag{3.29}$$

Additional constraints are introduced so that the thickness of the slab is not less than a lower bound, t_{cl}, and to ensure that all design variables are positive:

$$1 - \frac{t_c}{t_{cl}} \leq 0 \tag{3.30}$$

$$-\frac{A_s}{A_{si}} \leq 0 \tag{3.31}$$

$$-\frac{N_s}{N_{si}} \leq 0 \tag{3.32}$$

where $A_{\mathrm{si}} =$ the initial cross-sectional area of steel and $N_{\mathrm{si}} =$ the initial number of shear studs.

3.2.3 Problem Formulation as a Mixed Integer–Discrete Nonlinear Programming Problem

This optimization problem is characterized by only three design variables, t_{c}, N_{s}, and A_{s}, but highly complicated and discontinuous nonlinear constraints that are known to baffle optimization algorithms. The three variables are of different types. The number of shear studs is an integer. The thickness of the concrete slab can be treated as an integer by considering the fact that in practice the thickness has to be a round number, e.g. in centimeters in the SI system or a multiple of $\frac{1}{8}$ in. or $\frac{1}{4}$ in. in the US customary system of units. The cross-sectional area is a real number but has to be treated as a discrete variable because it has to be chosen from a limited number of commercially available sections, such as those given in the AISC LRFD manual (AISC, 2001). As such, the problem is formulated as a mixed integer–discrete nonlinear programming (MIDNLP) problem.

A continuous variable optimization problem can be defined as

$$\text{Minimize } f(\mathbf{x}) \text{ subject to } g_i(\mathbf{x}) \le 0, \quad i = 1, \ldots, j \tag{3.33}$$

where $\mathbf{x} =$ a real vector of design variables, $f(\mathbf{x}) =$ the cost function, $g_i(\mathbf{x}) =$ the ith inequality constraint, and $j =$ the total number of inequality constraints. The functions $f(\mathbf{x})$ and $g_i(\mathbf{x})$ are assumed to be differentiable. For an MIDNLP problem, the optimization problem can be expressed as

$$\text{Minimize } f(\mathbf{x}) \text{ subject to } g_i(\mathbf{x}) \le 0, \quad i = 1, \ldots, j, \quad x_k \in D_k,$$

$$D_k (d_{k1}, d_{k2}, \ldots, d_{kn^k}), \quad k = 1, \ldots, n_{\mathrm{d}} \tag{3.34}$$

where $n_{\mathrm{d}} =$ number of discrete design variables, $D_k =$ set of discrete values for the kth variable, $n^k =$ number of discrete values for the kth variable, and $d_{kl} =$ the lth discrete value for the kth variable. The number of discrete values for each variable can be different, and discrete values can include integer values. The discrete design space is disjoint and nonconvex, and the inequality constraints may not be active at the optimum point due to the fact that the constraint surface may not pass through the discrete optimum points. The branch and bound method and simulated annealing method have been used to solve MIDNLP problems. They are, however, known to be very time consuming (Arora et al., 1994).

3.3 Solution by the Floating-Point Genetic Algorithm

3.3.1 Binary Versus Floating-Point GA

Genetic algorithms are usually implemented by binary representations of parameters, which are simple to create and manipulate. The binary representation, however, has shortcomings when the number of variables is large or a high level of precision is required. The binary representation of parameters has limited precision. For a large number of variables, large computational processing resources would be required even for small sizes of population. An alternative to the popular binary genetic algorithm is the floating-point parameter GA, or floating-point GA for short, where a floating-point number represents each variable.

3.3.2 Crossover Operation for the Floating-Point GA

Consider two parent strings $P(n)_1$ and $P(n)_2$ for an N-variable problem:

$$P(n)_1 = [p_{11}, {}^\downarrow p_{12}, p_{13}, p_{14}, {}^\downarrow p_{15}, \ldots, p_{1N}] \tag{3.35}$$

$$P(n)_2 = [p_{21}, {}^\downarrow p_{22}, p_{23}, p_{24}, {}^\downarrow p_{25}, \ldots, p_{2N}] \tag{3.36}$$

where the floating-point values for variable are separated by commas. This is in contrast to the binary GA where the values are concatenated without any space or comma. In the case of the floating-point GA the crossover points can be between variables only. The vertical arrows in Eqs. (3.35) and (3.36) indicate the crossover points for the example shown.

The problem with such a crossover is that variables cannot receive new values in subsequent generations as initial randomly generated continuous variable values are simply propagated to the succeeding generations with only different combinations. To remedy this problem, linear interpolation is used to combine variable values with the crossovers from parents to an offspring in the following form (Michalewicz, 1995):

$$o_{1n} = \beta p_{1n} + (1 - \beta) p_{1n} \tag{3.37}$$

$$o_{2n} = \beta p_{2n} + (1 - \beta) p_{2n} \tag{3.38}$$

where β is a random number in the interval [0,1]. The two offsprings $O(n)_1$ and $O(n)_2$ become:

$$O(n)_1 = [p_{11}, {}^\downarrow o_{12}, o_{13}, o_{14}{}^\downarrow p_{15}, \ldots, p_{1N}] \tag{3.39}$$

$$O(n)_2 = [p_{21}, \,^{\downarrow}o_{22}, o_{23}, o_{24}, \,^{\downarrow}p_{25}, \dots, p_{2N}] \qquad (3.40)$$

When β is equal to 0.5, the results for both offsprings are the average values of the parent variables. However, limiting the value of β in the interval $[0,1]$ does not allow introduction of values beyond the extremes of the parent values in the population. By selecting a value of β greater than 1, it is possible to extrapolate beyond the extremes.

3.3.3 Mutation Operation for the Floating-Point GA

For the binary GA, changing a bit from 0 to 1 or vice versa performs a mutation. In the floating-point GA, a variable is selected randomly and exchanged with a new randomly generated floating point-number.

3.3.4 Floating-Point GA for Cost Optimization of Composite Floors

To solve the discrete optimization problem, a floating-point genetic algorithm is applied to obtain a continuous variable optimum solution. Next, this solution is mapped up or down to the nearest discrete solution, one at a time for each variable, until the discrete optimum solution is found, as described subsequently.

Following Schoenauer and Xanthakis (1993), the continuous variable optimum solution is found in two steps and using two different fitness functions. In the first step, the GA is used for constraint satisfaction only. The magnitude of the violation of the ith constraint is defined as

$$C_i(\mathbf{x}) = \max[g_i(\mathbf{x}), 0], \quad i = 1, \dots, m \qquad (3.41)$$

and the following fitness function is used in this step:

$$F_1 = M_i^n - C_i \qquad (3.42)$$

where

$$M_i^n = \max[C_i] \qquad (3.43)$$

Therefore, using the fitness function F_1, GA produces designs with smaller constraint violations in every generation. This fitness function also ensures all feasible designs have the same fitness, thus preventing convergence biased to a particular feasible design.

Infeasible points for the first to $(i-1)$st constraints are assigned null fitness and discarded in the following population selection step. Evolution for the ith constraint ends when a large enough part of the population (a preselected $\tau\%$) becomes feasible for that constraint. Consequently, $\tau\%$ of the population after m iterations of the constraint evolution is feasible, satisfying all the constraints.

In the next step of the cost optimization algorithm, an objective function is defined as a combination of the cost function, $f(\mathbf{x})$, and penalized violations of the constraints:

$$F_2 = f(\mathbf{x}) + \lambda \sum_{i=1}^{m} (\max[g_i(\mathbf{x}), 0]) \tag{3.44}$$

where $\lambda =$ the Lagrange multiplier. After the continuous optimization solution is found, the continuous optimum design values are mapped up or down to the nearest discrete variable one variable at a time. Then, following Ringertz (1988), the perturbation of the penalty function is computed as follows:

$$\Delta V = V\left(\mathbf{x}^j, \lambda\right) - V(\mathbf{x}, \lambda) \tag{3.45}$$

where

$$V(\mathbf{x}, \lambda) = f(\mathbf{x}) + \lambda \sum_{k=1}^{m} g_k(\mathbf{x}) \tag{3.46}$$

and $\mathbf{x}^j =$ vector of perturbed design variables for the jth mapped design variable. The variable that yields the minimum penalty is kept constant at its discrete value and the constrained floating-point GA is applied again to find updated values for the remaining design variable(s). This process is performed iteratively until all design variables are set to be discrete, as summarized below:

Step 1. Set the constraint counter, $i = 1$, and initialize the population.
Step 2. Set the generation counter, $n = 1$, and initialize the Lagrange multiplier, λ.
Step 3. Calculate $C_i(\mathbf{x})$, M_i^n, and assign fitnesses to individuals:

$$F_1 = \begin{cases} M_i^n - C_i & \text{if } C_1 = \cdots = C_{i-1} = 0 \\ 0 & \text{otherwise} \end{cases} \tag{3.47}$$

Step 4. Generate the next population. Evaluate the jth constraint, if $\tau\%$ of the population is feasible, go to step 5; otherwise update the Lagrange multiplier, set $n = n + 1$, and go to step 2.

Step 5. If $i = \mathrm{m}$, go to step 6; otherwise set $i = i + 1$ and go to step 1.

Step 6. Find the continuous variable optimum solution using the penalty function-based GA:

$$F_2 = f(\mathbf{x}) + \lambda \sum_{i=1}^{m} (\max\,[g_i(\mathbf{x}), 0]) \qquad (3.48)$$

Step 7. For $j = 1$ to n_d, calculate the perturbation of the penalty function:

$$\Delta V = \left[f(\mathbf{x}^j) + \lambda \sum_{i=1}^{m} g_i(\mathbf{x}^j) \right] - \left[f(\mathbf{x}) + \lambda \sum_{i=1}^{m} g_i(\mathbf{x}) \right] \qquad (3.49)$$

Select the design variables with the least ΔV and set $\mathbf{x}^j$, excluding the jth variable, as the new initial design variables. The jth variable is kept constant at its discrete value. Set $n_d = n_d - 1$.

Step 8. If all variables are set to be discrete, stop; otherwise, go to Step 1.

In steps 2 to 6, the counter propagation neural (CPN) network is used to map the continuous variable cross-sectional area to a discrete commercially available section (Hecht-Nielsen, 1987, 1988; Adeli and Park, 1998) (see Section 3.5). In step 7 the CPN network is used as a continuous function approximator.

3.4 Solution by the Neural Dynamics Method

Adeli and Park (1998) developed a neural dynamics model particularly suitable for large-scale and complicated optimization problems for which they received a US Patent on 29 September 1998 (US Patent 5,815,394). The stability and robustness of the model has been demonstrated by application to optimization of a number of very large structures including a 144-story steel building structure with more than 20 000 members subjected to actual loading and complicated highly nonlinear implicit and discontinuous constraints of the AISC LRFD code (AISC, 2001). The neural dynamics model has also been applied successfully to optimization of cold-formed steel beams (Adeli and Karim, 1997a; Karim and Adeli, 1999a, 1999b) and to scheduling and cost optimization of construction projects (Adeli and Karim, 1997b). In this section, the neural dynamics model of Adeli and Park (1998) is employed to solve the composite floor cost optimization problem.

To solve the MIDNLP problem, first the neural dynamics optimization model of Adeli and Park (1998) is applied, assuming all the variables are continuous. Next, the continuous optimum design values are mapped up or down to the nearest discrete variable. Then, following Ringertz (1988), the perturbation of the penalty function is computed as noted by equations (3.45) and (3.46). The variable that yields the minimum penalty is kept constant at its discrete value and the neural dynamics model is applied again to find updated values for the other two design variables. This process is performed iteratively until all design variables are set to be discrete using the following steps:

Step 1. Solve the continuous design optimization problem using the neural dynamics model of Adeli and Park (1998) (step 1 to step 5). Set initial design variables and an initial Lagrange multiplier, λ_0. Set the iteration counter $n = 1$.

Step 2. Calculate the gradients of the cost function, C, for each variable

$$C_i = -\frac{\partial F(\mathbf{x})}{\partial x_i}, \quad i = 1, \ldots, n_\mathrm{d} \tag{3.50}$$

Step 3. Update the Lagrange multiplier and calculate the design constraints. If constraints are violated, compute the gradients of the constraints:

$$\lambda_\mathrm{n} = \lambda_0 \sqrt{n^3} \tag{3.51}$$

$$O_j = \lambda_\mathrm{n} \max[0, g_j(\mathbf{x})], \quad j = 1, \ldots, m \tag{3.52}$$

$$Z_{ji} = \frac{\partial g_i(\mathbf{x})}{\partial x_i}, \quad j = 1, m; i = 1, \ldots, n_\mathrm{d} \tag{3.53}$$

where $O_j =$ the penalized output of the jth constraint and $Z_{ji} =$ the gradient of the jth constraints with respect to the ith variable. Equation (3.52) penalizes the constraint when it is not satisfied; otherwise, a zero is assigned to O_j.

Step 4. Update the design variables using the following learning rule of the neural dynamics model:

$$x_i^\mathrm{new} = x_i^\mathrm{old} + \int \left(C_i + \sum_{j=1}^{m} Z_{ji} O_j \right) \mathrm{d}t, \quad i = 1, n_\mathrm{d} \tag{3.54}$$

The Euler method is used to evaluate the integral.

Step 5. If the updated continuous variables are optimum, go to step 6; otherwise, set $n = n + 1$ and go to step 1.

Step 6. From $i = 1$ to $i = n_d$, calculate the perturbation of the penalty function:

$$\Delta V = \left[f(\mathbf{x}^i) + \lambda_n \sum_{k=1}^{m} g_k(\mathbf{x}^i) \right] - \left[f(\mathbf{x}) + \lambda_n \sum_{k=1}^{m} g_k(\mathbf{x}) \right] \quad (3.55)$$

Select the design variables, $\mathbf{x}^i$, with the least ΔV and set $\mathbf{x}^i$ excluding the ith variable as the new initial design variables for the neural dynamics model. The ith variable is kept constant at its discrete value. Set $n_d = n_d - 1$.

Step 7. If all variables are set to be discrete, stop; otherwise, go to step 1.

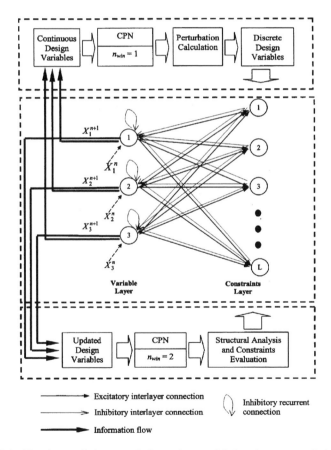

Figure 3.4 Topology of the neural dynamics model for the cost optimization of composite floors

The topology of the neural network model for cost optimization of composite floors is shown in Figure 3.4. Gradients of the cost function are computed explicitly and assigned to the inhibitory recurrent connections in the variable layer (equation (3.50)). These gradients provide the steepest direction to the optimal point. Gradient information of the constraints is assigned as weights of the inhibitory links connecting the variable layer to the constraints layer (equations (3.52) and (3.53)). Each node in the constraint layer represents a normalized constraint described in a previous section. The integral values of the sums of the penalized outputs times the constraint gradients and the negative of the gradient of the cost function become the updated input to the nodes in the input layer (equation (3.54)).

3.5 Counter Propagation Neural (CPN) Network for Function Approximations

In the formulation of the cost optimization of composite floors, a steel beam is defined by one variable, its cross-sectional area. In evaluating the composite floors design constraints, however, other design parameters come into play. They are d, b_f, t_f, t_w, I_s, and the plastic modulus, Z_s. Bhatti (1996) used the least squares fit to express the moment of inertia (I_s) and the plastic modulus (Z_s) in terms of the cross-sectional area (A_s) and the depth of steel beam (d), and reported errors in the order of 5 % for W12 to W30 shapes tabulated in the AISC LRFD manual (AISC, 2001).

The CPN network functions as a statistically self-learning look-up table. Hecht-Nielsen (1987, 1988) showed that the CPN network can produce a near-optimal mapping approximation for problems described by continuous variables. Adeli and Park (1998) showed that the CPN network can be used as a powerful and accurate tool for presenting the relationship among various discrete design variables. They reported very small error in the order of 0.1 %. Consequently, the CPN network is used for both discrete variable mapping to relate the cross-sectional properties of steel beams to the design variable (A_s) and to approximate these relations for the continuous variable phase of the algorithm.

The topology of the CPN network consists of a fully connected network of input, competition, and interpolation layers (Figure 3.5). During the training phase of the CPN network a winning node is selected in the competition layer based on the smallest Euclidean distance or dot product of the input vector and the vectors of the weights of links connecting the input and competition layers. This winner-takes-all operation in the competition layer is based on

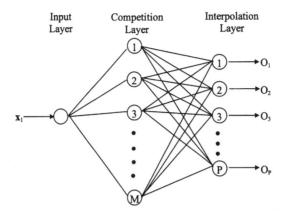

Figure 3.5 Topology of CPN

the Kohonen's unsupervised self-organizing learning rule (Kohonen, 1988). This operation yields the weights of the links connecting the input layer to the competition layer. Next, Grossberg's supervised learning rule (Grossberg, 1982) is used to find the weights of the links connecting the competition layer to the interpolation (output) layer.

The input layer has only one node representing the cross-sectional area of the steel beam. The number of nodes in the output layer, P, is equal to the number of cross-section properties of the steel beam which is equal to 7 (A_s, d, t_f, b_f, t_w, I_s, Z_s). The number of nodes in the hidden competition layer is chosen equal to the number of training instances. The properties of 77 W shapes commonly used in composite floors are given in the AISC LRFD manual (AISC, 2001). Therefore, the number of training instances is equal to 77.

The trained CPN network is used in steps 3 and 6 of the algorithm described in the previous section. In step 3, where the CPN network is used to map the continuous variable cross-sectional area to a discrete commercially available section, the number of winning nodes is just one. In step 6, where the CPN network is used as a continuous function approximator, several nodes are selected in the competition layer in the neighborhood of the winning node on the basis of the shortest Euclidean distance and the following output vector is calculated for the selected winning nodes:

$$\mathbf{O} = \sum_{j=1}^{n_{\text{win}}} \left\{ \frac{\left[\left(\sum_{k=1}^{n_{\text{win}}} d_k \right) - d_j \right]}{(n_{\text{win}} - 1) \sum_{k=1}^{n_{\text{win}}} d_k} \mathbf{O}_j \right\} \quad \text{when } n_{\text{win}} \geq 2 \tag{3.56}$$

where n_{win} = number of nodes selected in the neighborhood of the winning node, d_k = Euclidean distance between the input vector and the weight vector of links connecting the input layer to the kth winning node, and $\mathbf{O}_j =$ the output vector associated with the jth winning node in the competition layer. The operand inside the outer Σ is adjusted so that the sum of the coefficients of all the output vectors is equal to one. Values of 2, 3, and 4 were investigated for n_{win}. It was found that $n_{win} = 2$ produces satisfactory results.

Figures 3.6 and 3.7 show a comparison of results obtained from the CPN network and the exact values from the AISC manual for the moment of inertia and the depth, respectively. Two important conclusions are made. First, the relationships are complicated and cannot be represented accurately by traditional statistical curve-fitting approaches. Second, the CPN network can represent the mapping with excellent accuracy of less than 0.1 % for discrete values and 1 % for continuous values, in both cases indistinguishable in Figures 3.6 and 3.7. The larger error value for the case of continuous values is due to the fact that another level of approximation is involved. In this case the output of the CPN has to be interpolated over the winning node and its closest node (in terms of the Euclidean distance) linearly, which creates the additional approximation. Results similar to those shown in Figures 3.6 and 3.7 have been found for the other cross-sectional properties $(A_s, t_f, b_f, t_w, Z_s)$.

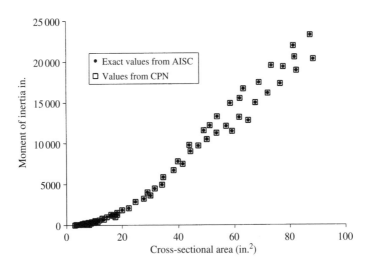

Figure 3.6 Relationship between the cross-sectional area and the moment of inertia, I_s

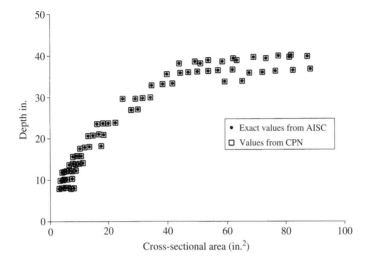

Figure 3.7   Relationship between the cross-sectional area and the depth of steel

3.6 Examples

3.6.1 Example 1

This example is the optimum cost design of a rectangular floor area with input design values summarized in Table 3.1. The data are taken from Salmon and Johnson (1996) for the sake of comparison. They chose a concrete slab thickness of $t_c = 5$ in. (127 mm). The floor is constructed without shores and $\frac{3}{4}$ in. diameter × 3 in. (19 mm diameter × 76 mm) shear studs are used.

As Lorenz (1988) points out, the relative cost coefficients C_{sc} and C_{ss} can vary based on the job size and the geographical regions. He suggests a range of 6 to 12 for C_{sc} but presents none for C_{ss} as he does not consider the cost of concrete. For this particular example, values of $C_{sc} = 3$ and $C_{ss} = 10$

Table 3.1   Input design parameters for the composite floor design of Example 1

F_y	50 ksi (344.7 MPa)
f'_c	3 ksi (20.7 MPa)
B	8 ft (2.44 m)
γ_c	150 lb/ft^3 (23.56 kN/m^3)
L	30 ft (9.15 m)
Live load	150 psf (7.2 kPa)

are chosen for the cost coefficients based on Means (1999). The live load deflection is limited to $L/360$.

Like Salmon and Johnson (1996), the concrete slab thickness is kept constant and is equal to 5 in. ($t_c = 5$ in.). The cost optimization algorithm yielded a solution of $N_s = 20$ and $A_s = 9.12$ in.^{2} (W16 × 31), which happened to be the same as that given in Salmon and Johnson (1996). This can be explained by the fact that this is the minimum number of shear studs as given by the AISC LRFD manual and governs the solution for this example.

Next, the concrete slab thickness was considered as a design variable with a lower bound of 4 in (102 mm). Three different sets of initial design values were used to study the convergence of the cost optimization algorithm. The results of optimization are shown in Table 3.2 and Figure 3.8. The optimum cost solution obtained in this research is about 9 % less expensive than that presented in Salmon and Johnson (1996).

3.6.2 Example 2

This example is the optimum cost design of a rectangular floor area with input design values summarized in Table 3.3. The data are taken from Salmon and Johnson (1996) for the sake of comparison. They chose a concrete slab thickness of $t_c = 4$ in. (101.6 mm). The floor is constructed without shores and $\frac{3}{4}$ in. diameter × 3 in. (19 mm diameter × 76 mm) shear studs are used. Values of $C_{sc} = 3$ and $C_{ss} = 12$ are chosen for the cost coefficients.

Table 3.2   Results of the cost optimization for Example 1

Case number	t_c (in.)	N_s	A_s (in.^{2})	Total relative cost
Set A				
Starting point	35.0	35	40.00	5502
Discrete optimum	4.0	28	7.68(W16 × 26)	1304
Set B				
Starting point	25.0	25	30.00	3980
Discrete optimum	4.0	28	7.68(W16 × 26)	1304
Set C				
Starting point	15.0	17	20.00	2620
Discrete optimum	4.0	28	7.68(W16 × 26)	1304
Salmon and Johnson (1996)	5.0	20	9.12(W16 × 31)	1431

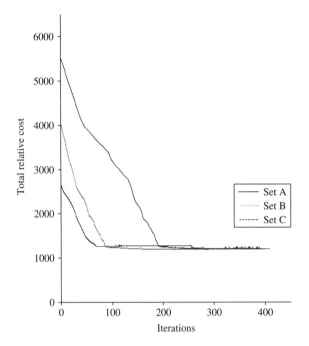

Figure 3.8 Convergence history for Example 1 using different sets of starting points, as defined in Table 3.2

Table 3.3 Input design parameters for the composite floor design of Example 2

F_y	36 ksi (217.7 MPa)
f'_c	3 ksi (20.7 MPa)
B	8 ft (2.44 m)
γ_c	150 lb/ft^3 (23.56 kN/m^3)
L	28 ft (8.54 m)
Live load	150 psf (7.2 kPa)

For the constant thickness of concrete, $t_c = 4$ in., the same as Salmon and Johnson (1996), the cost optimization algorithm yielded a solution of $N_s = 24$ and $A_s = 9.12$ in.^{2} (W16 × 31) with a relative total cost of 1381, resulting in about 6.5 % of savings compared with the solution given in Salmon and Johnson (1996).

With a lower bound of 3.5 in. (88.9 mm) and an increment of $\frac{1}{8}$ in. for the slab thickness, three different sets of initial design values were used. The results of the optimization are shown in Table 3.4 and Figure 3.9. The

Table 3.4 Results of the cost optimization for Example 2

Case number	t_c (in.)	N_s	A_s (in.^{2})	Total relative cost
Set A				
Starting point	25.0	60	50.00	6884
Discrete optimum	3.875	24	9.12 (W16 × 31)	1373
Set B				
Starting point	15.0	50	40.00	5251
Discrete optimum	3.875	24	9.12 (W16 × 31)	1373
Set C				
Starting point	10.0	30	30.00	2826
Discrete optimum	3.875	24	9.12 (W16 × 31)	1373
Salmon and Johnson (1996)	4.0	32	9.12 (W16 × 31)	1477

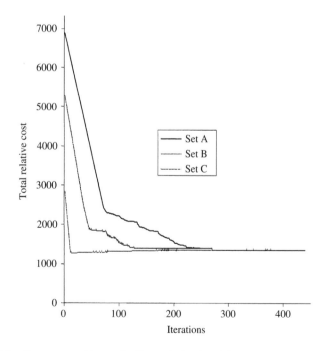

Figure 3.9 Convergence history for Example 2 using different sets of starting points as defined in Table 3.4

optimum cost solution obtained in this case is about 7 % less expensive than that presented in Salmon and Johnson (1996). The CPU (central processing unit) times required to find the optimum solution for both examples are in the range of 10 s to 20 s on a 400 MHz personal computer.

4

Fuzzy Genetic Algorithm for Optimization of Steel Structures

4.1 Introduction

A goal of the authors' research is to model the effects of imprecision, uncertainty, or fuzziness in the formulation of a GA-based structural design optimization problem. Soh and Yang (1996) discuss a fuzzy controlled genetic search algorithm for the minimum weight shape optimization of steel trusses by coupling a heuristic fuzzy rule-based system (Adeli, 1988, 1990) with the GA without using the actual constraints of any commonly used design code. The heuristic fuzzy rules are intended to improve the performance of GA by (a) introducing experiential rules acquired from experts, (b) guiding the search for an optimum shape, and (c) employing the fuzzy representation of the design variables. The primary focus of their work is to improve the optimization process through the use of heuristic rules.

In a highly constrained optimization problem such as structural optimization, maintaining feasibility can be a bottleneck for genetic search and hence some sort of relaxation of constraints in the early stage of search is recommended (Smith and Tate, 1993). In this chapter, a fuzzy genetic algorithm is formulated for global optimization of steel structures subjected to the

Cost Optimization of Structures: Fuzzy Logic, Genetic Algorithms, and Parallel Computing   H. Adeli and K. C. Sarma   © 2006 John Wiley & Sons, Ltd

constraints of the AISC specifications taking into account the imprecision and fuzziness in the constraints. If a candidate solution is discarded in the early stages of the genetic search because of a small violation of one or more constraints, then the search may miss the potentially global optimum in the vicinity of the candidate solution. In other words, by treating the constraints as fuzzy constraints the chance of obtaining the global optimum will be increased.

A lot has been said in the literature about the virtues of GAs as not being entrapped in a local optimum and its ability to yield the global optimum solution. However, to find the global optimum for large problems with hundreds of design variables, large population sizes with a large number of iterations are needed, resulting in prohibitively large function evaluations in each iteration and an enormous computer processing time which may not be readily possible even on a supercomputer. This limitation of GAs for large-scale design optimization problems is hardly discussed in the growing GA literature. Furthermore, while GAs can identify the global region they may have difficulty converging to the exact value of the global optimum. This convergence difficulty is in fact noted in the recent literature for heavily constrained optimization problems such as structural optimization, where the feasible design space is usually nonconvex and consists of small noncontiguous regions (Powell and Skolnick, 1993; Schoenauer and Xanthakis, 1993; Kim and Myung, 1996, 1997).

The simple one-dimensional unconstrained minimization problem shown in Figure 4.1 explains this point in an approximate manner. A simple GA identifies the region of the global optimum and yields a solution within range B in the vicinity of the global optimum. However, the solution then often oscillates in this range or a very large number of iterations is required in order to obtain the exact value of the global optimum, noted by W^* in Figure 4.1. It must be pointed out that the simple example of Figure 4.1 is used only to illustrate the point; the convergence problem of GAs actually occurs in highly constrained nonconvex optimization problems.

Another objective of the authors' research is to improve the convergence and efficiency of GAs through the use of fuzzy set theory. Our fuzzy GA consists of two steps. In the first step the augmented Lagrangian GA of Adeli and Cheng (1994a), described in Sections 2.5 and 2.6, is used to reach the region of the global optimum within relatively few iterations. Then, the solution obtained in the first step is improved through a local fuzzy GA search by including rules based on the fuzzy set theory and appropriate fuzzy membership functions.

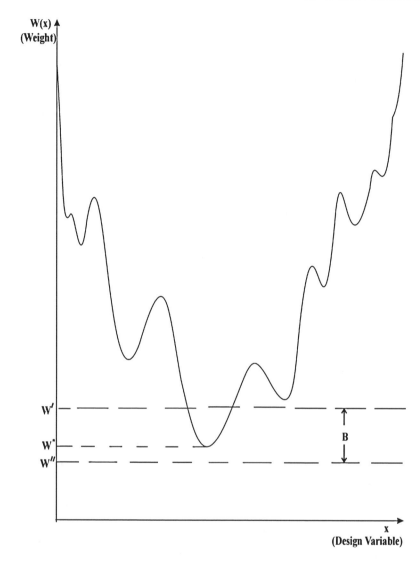

Figure 4.1 An example of a one-dimensional unconstrained minimization problem

4.2 Fuzzy Set Theory and Structural Optimization

The theory of fuzzy sets, developed by Zadeh (1965), can be used to model imprecision, ambiguity, or fuzziness in the formulation. In the formulation of the structural optimization problem a major source of imprecision or fuzziness exists in the evaluation of the constraints. In traditional optimization algorithms, constraints are satisfied within a tolerance defined by a crisp or nonfuzzy number. In reality and engineering practice this evaluation involves many sources of approximation. A design is commonly considered satisfactory when the constraints are satisfied within a given predetermined tolerance such as 1 %. When an optimization algorithm satisfies the constraints exactly (within the small tolerance of numerical computations) it can miss the true optimum solution within the confine of practical and realistic approximations. It will be demonstrated that by taking into account the fuzziness and imprecision in the constraints and employing the fuzzy set theory it is possible to reduce the objective function further and substantially increase the probability of finding the actual global optimum solution.

Fuzziness is considered both in the constraints and the objective function. Several articles have been published on fuzzy optimization of structures (Wang and Wang, 1985a, 1985b; Rao, 1987a, 1987b; Yeh and Hsu, 1990). With the exception of Soh and Yang (1996) mentioned earlier, however, all of them use conventional nonevolutionary optimization algorithms.

A crisp nonfuzzy structural optimization is formulated as: find the vector of design variables $\mathbf{x}$ such that the weight of the structure $W(\mathbf{x})$ is minimized subject to the equality and inequality constraints:

$$h_i(\mathbf{x}) = 0, \quad i = 1, 2, \ldots, N_{ch} \tag{4.1}$$

$$g_i^l(\mathbf{x}) \le g_i(\mathbf{x}) \le g_i^u(\mathbf{x}), \quad i = 1, 2, \ldots, N_{cg} \tag{4.2}$$

where N_{ch} and N_{cg} are the total number of equality and inequality constraints, respectively; $g_i^u(\mathbf{x})$ and $g_i^l(\mathbf{x})$ are the upper and lower bounds on the constraint $g_i(\mathbf{x})$.

When fuzziness is considered in both the objective function and constraints, the design variables $(\mathbf{x})$ can be obtained from a fuzzy domain D such that the membership function μ_D for the fuzzy domain D can be obtained from the intersection of the fuzzy membership functions for both the objective function and constraints as follows:

$$\mu_D = \mu_W(\mathbf{x}) \cap \left[\bigcap_{i=1,2,\ldots,N_{cg}} \mu_{g_i}(\mathbf{x}) \right] \tag{4.3}$$

where $\mu_W(\mathbf{x})$ and $\mu_{g_i}(\mathbf{x})$ are the membership functions for the objective function and the ith inequality design constraint, respectively. From this fuzzy domain, the optimum solution $(\mathbf{x}^*)$ for the design variable $\mathbf{x}$ can be obtained by using the following max-min procedure (Bellman and Zadeh, 1970):

$$\mu_D(\mathbf{x}^*) = \text{maximize } \mu_D(\mathbf{x}) \tag{4.4}$$

where

$$\mu_D(\mathbf{x}) = \min\left[\mu_W(\mathbf{x}), \min_{i=1,2,\ldots,N_{cg}} \mu_{g_i}(\mathbf{x})\right] \tag{4.5}$$

This max-min procedure can be solved by maximizing a scalar parameter λ, known as the overall satisfaction parameter, in the following way (Zimmermann, 1978). Find the vector of design variables $\mathbf{x}$ and the parameter λ such that:

Maximize λ subject to the constraints:

$$\lambda \leq \mu_W(\mathbf{x}) \tag{4.6}$$

$$\lambda \leq \mu_{g_i}^u(\mathbf{x}), \quad i = 1, 2, \ldots, N_{cg} \tag{4.7}$$

$$\lambda \leq \mu_{g_i}^l(\mathbf{x}), \quad i = 1, 2, \ldots, N_{cg} \tag{4.8}$$

$$0 \leq \lambda \leq 1 \tag{4.9}$$

where $\mu_{g_i}^u(\mathbf{x})$ and $\mu_{g_i}^l(\mathbf{x})$ are the membership functions for the upper and lower bounds of the inequality constraints $g_i(\mathbf{x})$ (equation (4.2)), respectively. The equality constraints, in equation (4.1) are not included in the fuzzy formulations because they have to be satisfied strictly.

In this chapter two sets of fuzzy inequality constraints are considered: one for the displacements and the other for the stresses. Thus, equations (4.7) and (4.8) are rewritten as

$$\lambda \leq \mu_{\delta_i}^u(\mathbf{x}), \quad i = 1, 2, \ldots, N_d \tag{4.10}$$

$$\lambda \leq \mu_{\delta_i}^l(\mathbf{x}), \quad i = 1, 2, \ldots, N_d \tag{4.11}$$

$$\lambda \leq \mu_{\sigma_i}^u(\mathbf{x}), \quad i = 1, 2, \ldots, N_e \tag{4.12}$$

$$\lambda \leq \mu_{\sigma_i}^l(\mathbf{x}), \quad i = 1, 2, \ldots, N_e \tag{4.13}$$

where $\mu_{\delta_i}^u(\mathbf{x})$, $\mu_{\delta_i}^l(\mathbf{x})$, $\mu_{\sigma_i}^u(\mathbf{x})$, and $\mu_{\sigma_i}^l(\mathbf{x})$ are the membership functions for the upper and lower bounds on displacements and stresses, respectively; N_d is the number of constrained degrees of freedom and N_e is the number of elements.

4.3 Minimum Weight Design of Axially Loaded Space Structures

In this chapter attention is limited to structures consisting of axially loaded members subjected to the constraints of the AISC ASD code (AISC, 1995). The nonfuzzy minimum weight design of such structures can be formulated as

$$\text{Minimize} \quad W(\mathbf{x}) = \rho \, \mathbf{L}'\mathbf{x} \tag{4.14}$$

subject to the constraints:

$$\boldsymbol{\delta} \leq \boldsymbol{\delta}^{a} \tag{4.15}$$

$$\boldsymbol{\sigma} \leq \boldsymbol{\sigma}^{a} \tag{4.16}$$

$$\mathbf{x}^{l} \leq \mathbf{x} \leq \mathbf{x}^{u} \tag{4.17}$$

where ρ is the unit weight of materials; $\mathbf{L}$ and $\mathbf{x}$ are the vectors of lengths and cross-sectional areas of the groups of members in the structure, respectively; prime indicates the transpose of a matrix; $\boldsymbol{\delta}$ is the vector of nodal displacement degrees of freedom; $\boldsymbol{\sigma}$ is the vector of stresses in the elements of the structure; and superscripts l, u, and a refer to lower and upper bounds, and allowable values, respectively. The allowable displacements (δ_i^a) can be expressed as

$$\delta_i^a = \begin{cases} -\delta_i^l & \text{for } \delta_i < 0, \\ \delta_i^u & \text{for } \delta_i \geq 0, \end{cases} \quad \text{for } i = 1, 2, \ldots, N_d \tag{4.18}$$

Similarly, the allowable stresses (σ_i^a) can be expressed as

$$\sigma_i^a = \begin{cases} -\sigma_i^l & \text{for } \sigma_i < 0, \\ \sigma_i^u & \text{for } \sigma_i \geq 0, \end{cases} \quad \text{for } i = 1, 2, \ldots, N_e \tag{4.19}$$

According to the AISC ASD code (AISC, 1995) the allowable tensile and compressive stresses in the ith element are calculated as

$$\sigma_i^u = 0.6F_y \tag{4.20}$$

and

$$\sigma_i^l = \begin{cases} \dfrac{\left(1 - \dfrac{\zeta_i^2}{2C_c^2}\right) F_y}{\dfrac{5}{3} + \dfrac{3\zeta_i}{8C_c} - \dfrac{\zeta_i^3}{8C_c^3}}, & \text{for } \zeta_i < C_c \\[6mm] \dfrac{12\pi^2 E}{23\zeta_i^2}, & \text{for } \zeta_i \geq C_c \end{cases} \tag{4.21}$$

where ζ_i is the slenderness ratio, defined as

$$\zeta_i = \frac{k_i l_i}{r_i}, \quad i = 1, 2, \ldots, N_e \tag{4.22}$$

and k_i is the effective length factor and l_i and r_i are the length and the minimum radius of gyration of the ith element, respectively; F_y is the yield stress of steel; E is the modulus of elasticity; and C_c is the slenderness ratio dividing the elastic and inelastic buckling ranges, expressed as

$$C_c = \sqrt{2\pi^2 E / F_y} \tag{4.23}$$

The value of limiting compressive stress (equation (4.21)) depends on the minimum radius of gyration that can be considered as a second design variable for each member. This will double the size of the optimization problem and increase the computational cost several times. For the sake of computational efficiency, the radius of gyration of the cross-section is related to its cross-sectional area (the design variable) using a piecewise regression analysis. Adeli and Balasubramanyam (1988) used a piecewise linear regression analysis. In the present work, a piecewise parabolic relationship is employed between the radius of gyration, r, and the cross-sectional area, x, in the following form:

$$r = a_1 x^2 + a_2 x + a_3 \tag{4.24}$$

The most commonly used W shapes for axially loaded members are W8, W10, W12, and W14. In this work, thirteen W8, eighteen W10, twenty-nine

W12, and thirty-six W14 shapes are used as given in the AISC ASD manual (AISC, 1995). Figure 4.2 shows the actual data (minimum radius of gyration versus the cross-sectional area) for the thirty-six W14 shapes along with the computed data using the piecewise parabolic relationship. In order to obtain a close fit, the data in the figure have been divided into six groups, identified by circled numbers, and a parabolic regression analysis is performed for each group. This piecewise parabolic curve-fitting results in a maximum error of 0.2 %. Similar curve-fitting is performed for W8, W10, and W12 sections.

Any combination of W8, W10, W12, and W14 can be used in space truss axial load structures. In every iteration of the optimization process for any value obtained for a design variable (cross-sectional area) four corresponding minimum and maximum radii of gyration are computed for four different groups of W shapes using the piecewise parabolic relationships. Then, the ratio of minimum to maximum radii of gyration is calculated for each group

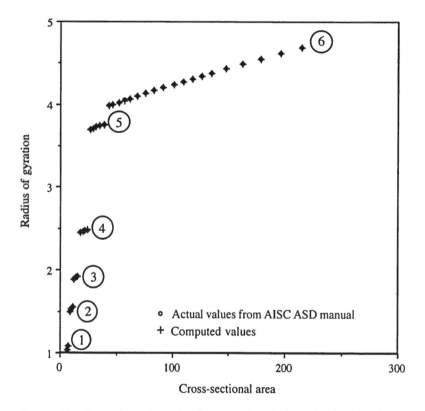

Figure 4.2 Comparison of actual and computed radii of gyration for W14 shapes

of W shapes; the value nearest to one will be the basis for selection of the minimum radius of gyration to be used in the stress constraint. This scheme is intended to yield economical selections as far as the buckling of compression members is concerned.

4.4 Fuzzy Membership Functions

In the present fuzzy GA optimization model, nonlinear quadratic fuzzy membership functions are used for both the objective function and constraints. The nonlinear fuzzy membership function for the objective function is presented as

$$\mu_W(x) = \begin{cases} 1, & \text{if } W(\mathbf{x}) \leq W''(\mathbf{x}) \\ b_1 W(\mathbf{x})^2 + b_2 W(\mathbf{x}) + b_3, & \text{if } W''(\mathbf{x}) < W(\mathbf{x}) < W'(\mathbf{x}) \\ 0, & \text{if } W(\mathbf{x}) \geq W'(\mathbf{x}) \end{cases} \quad (4.25)$$

In this equation W' is the weight obtained after several iterations of the simple non-fuzzy GA for structural optimization and W'' is an estimated lower bound for the global optimum W^*. The quantities W^*, W', and W'' are shown in Figure 4.1 for a one-dimensional unconstrained optimization problem. It is also possible to skip the simple GA and go directly to the fuzzy GA if W' can be estimated, say through design experience. The lower bound for the true global optimum, W'', can be estimated by reducing W' by a certain percentage.

The value of W' shown in Figure 4.1 is either a local optimum with no constraint violations or close to a local optimum without any or with small constraint violations. It is also possible for W' (say, obtained after several iterations of the simple GA) to be smaller than the value of the global optimum W^*, in which case it represents only an initial solution from the infeasible region with constraint violations. The present fuzzy GA can handle this situation as well. In this case, W'' becomes the upper bound, which is estimated by increasing W' by a certain percentage. The same equation (4.25) can be used by simply swapping W' and W''.

The values of the constants b_1, b_2, and b_3 in equation (4.24) are found by solving the following three simultaneous equations (Dhingra *et al.*, 1992):

$$b_1 \left\{ W''(\mathbf{x}) \right\}^2 + b_2 \left\{ W''(\mathbf{x}) \right\} + b_3 = 1.0 \qquad (4.26)$$

$$b_1 \left\{ W'(\mathbf{x}) \right\}^2 + b_2 W'(\mathbf{x}) + b_3 = 0.0 \qquad (4.27)$$

$$b_1 \left\{ W_{\text{ave}}(\mathbf{x}) \right\}^2 + b_2 W_{\text{ave}}(\mathbf{x}) + b_3 = 0.5 \qquad (4.28)$$

where

$$W_{ave}(\mathbf{x}) = \frac{W'(\mathbf{x}) + W''(\mathbf{x})}{2}$$

A graphical representation of the fuzzy membership function for the objective function is presented in Figure 4.3.

Similarly, the nonlinear fuzzy membership functions for upper bounds on constraints (g_i^u) are presented as follows:

$$\mu_{gi}^u(\mathbf{x}) = \begin{cases} 1, & \text{if } g_i(\mathbf{x}) \leq g_i^u(\mathbf{x}) \\ c_1 g_i(\mathbf{x})^2 + c_2 g_i(\mathbf{x}) + c_3, & \text{if } g_i^u(\mathbf{x}) < g_i(\mathbf{x}) < g_i^u(\mathbf{x}) + \Delta g_i^u(\mathbf{x}) \\ 0, & \text{if } g_i(\mathbf{x}) \geq g_i^u(\mathbf{x}) + \Delta g_i^u(\mathbf{x}) \end{cases}$$

$$(4.29)$$

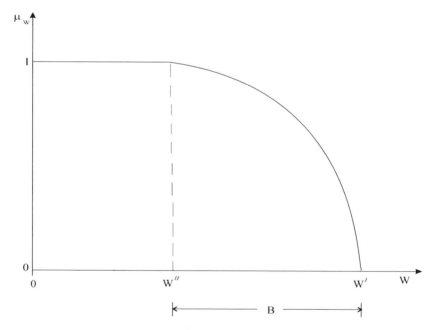

Figure 4.3 Graphical representation of the fuzzy membership function for the objective function

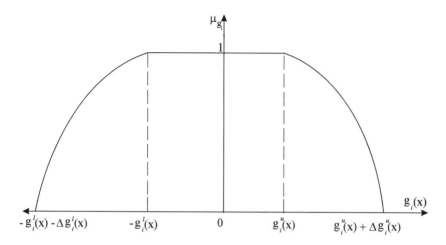

Figure 4.4 Graphical representation of the fuzzy membership function for upper and lower bounds on the constraints

where the values of the constants c_1, c_2, and c_3 are obtained by solving the following set of quadratic equations (Dhingra *et al.*, 1992):

$$c_1 \left\{ g_i^u(\mathbf{x}) \right\}^2 + c_2 g_i^u(\mathbf{x}) + c_3 = 1.0$$

$$c_1 \left\{ g_i^u(\mathbf{x}) + \Delta g_i^u(\mathbf{x}) \right\}^2 + c_2 \left\{ g_i^u(\mathbf{x}) + \Delta g_i^u(\mathbf{x}) \right\} + c_3 = 0.0$$

$$c_1 \left\{ g_{i\,\text{ave}}^u(\mathbf{x}) \right\}^2 + c_2 g_{i\,\text{ave}}^u(\mathbf{x}) + c_3 = 0.5 \qquad (4.30)$$

$$g_{i\,\text{ave}}^u(\mathbf{x}) = \frac{g_i^u(\mathbf{x}) + g_i^u(\mathbf{x}) + \Delta g_i^u(\mathbf{x})}{2}$$

where $\Delta g_i^u(\mathbf{x})$ is the magnitude of the relaxation of the upper bound for constraint i. The nonlinear fuzzy membership functions for the lower bounds on constraints, g_i^l, are expressed in a similar fashion. Figure 4.4 shows a graphical representation of the fuzzy membership functions for both upper and lower bounds on the constraints.

4.5 Fuzzy Augmented Lagrangian Genetic Algorithm

Genetic algorithms can be used directly to solve unconstrained optimization problems. In order to solve a constrained optimization problem by a GA, the constrained optimization problem has to be transformed to an unconstrained optimization problem. This can be done using the penalty function method or the augmented Lagrangian approach as discussed in Chapter 2. The advantages of the latter approach were noted in Chapter 2. In this section a fuzzy augmented Lagrangian GA is presented for optimization of structures subjected to multiple loading cases.

In the augmented Lagrangian GA, a Lagrangian functional is minimized in the following form:

$$
\text{Min } \varphi(\mathbf{x}) = \frac{1}{L_f} \sum_{i=1}^{N_e} \rho L_i x_i
$$

$$
+ \frac{1}{2} \left\{ \sum_{i=1}^{N_L} \sum_{j=1}^{N_d} \gamma_{i,j} \left[\left(\frac{|\delta_{i,j}|}{|\delta_j^a|} - 1 + \omega_{i,j} \right)^+ \right]^2 \right\}
$$

$$
+ \frac{1}{2} \left\{ \sum_{i=1}^{N_L} \sum_{j=1}^{N_e} \gamma_{i,j+N_d} \left[\left(\frac{|\sigma_{i,j}|}{|\sigma_j^a|} - 1 + \omega_{i,j+N_d} \right)^+ \right]^2 \right\} \qquad (4.31)
$$

where $\gamma_{i,j}$ = the Lagrangian penalty function parameter for the ith loading condition and the jth displacement or stress constraint, $\omega_{i,j}$ = the Lagrangian constraint parameter for the ith loading condition and the jth constraint, L_f = a factor for normalizing the objective function, and

$$
\left(\frac{|\delta_{i,j}|}{|\delta_j^a|} - 1 + \omega_{i,j} \right)^+ = \text{Max} \left(\frac{|\delta_{i,j}|}{|\delta_j^a|} - 1 + \omega_{i,j}, 0 \right) \qquad (4.32)
$$

$$
\left(\frac{|\sigma_{i,j}|}{|\sigma_j^a|} - 1 + \omega_{i,j} \right)^+ = \text{Max} \left(\frac{|\sigma_{i,j}|}{|\sigma_j^a|} - 1 + \omega_{i,j}, 0 \right) \qquad (4.33)
$$

Now, a fuzzy augmented Lagrangian functional is defined as a function of the design variable $\mathbf{x}$ and the overall satisfaction parameter λ, defined in a previous section (equations (4.6) to (4.13) as follows:

$$
\text{Min } \varphi(\mathbf{x}, \lambda) = - L_s \lambda + 1/2 \left\{ \alpha \left[\left(\frac{\lambda}{\mu_W} - 1 + \beta \right)^+ \right]^2 \right\}
$$

$$
+ \frac{1}{2} \left\{ \sum_{i=1}^{N_L} \sum_{j=1}^{N_d} \xi_{i,j} \left[\left(\frac{\lambda}{\mu_{\delta_{i,j}}^a} - 1 + \eta_{i,j} \right)^+ \right]^2 \right\}
$$

$$
+ \frac{1}{2} \left\{ \sum_{i=1}^{N_L} \sum_{k=1}^{N_e} \xi_{i,k+N_d} \left[\left(\frac{\lambda}{\mu_{\sigma_{i,k}}^a} - 1 + \eta_{i,k+N_d} \right)^+ \right]^2 \right\} \qquad (4.34)
$$

In this formulation the goal is to maximize the overall satisfaction factor, λ, subject to the constraints (equations (4.10) to (4.13)) to ensure that the satisfaction factor never exceeds the value of any membership function. In equation (4.34), α and β are the Lagrangian penalty function and constraint parameters, respectively, for the membership function of the objective function (weight of the structure). Similarly $\xi_{i,j}$ and $\eta_{i,j}$ are the Lagrangian penalty function and constraint parameters for the membership functions for the jth constraint (displacement and stress) due to the ith loading condition, $N_L =$ the number of loading cases, $L_s =$ a scaling factor, and $\mu^a_{\delta_{i,j}}$ and $\mu^a_{\sigma_{i,k}}$ are the membership functions for lower and upper bounds of the jth displacement and kth stress constraints for the ith loading condition, respectively, and are defined as follows:

$$\mu^a_{\delta_{i,j}} = \begin{cases} \mu^l_{\delta_{i,j}} & \text{for } \delta_{i,j} < 0, \\ \mu^u_{\delta_{i,j}} & \text{for } \delta_{i,j} \geq 0, \end{cases} \quad i = 1, 2, \ldots, N_L; j = 1, 2, \ldots, N_d \quad (4.35)$$

$$\mu^a_{\sigma_{i,k}} = \begin{cases} \mu^l_{\sigma_{i,k}} & \text{for } \sigma_{i,k} < 0, \\ \mu^u_{\sigma_{i,k}} & \text{for } \sigma_{i,k} \geq 0, \end{cases} \quad i = 1, 2, \ldots, N_L; k = 1, 2, \ldots, N_e$$

$$(4.36)$$

The terms inside the parentheses in equation (4.34) are defined as follows:

$$\left(\frac{\lambda}{\mu_W} - 1 + \beta \right)^+ = \text{Max}\left(\frac{\lambda}{\mu_W} - 1 + \beta, 0 \right) \quad (4.37)$$

$$\left(\frac{\lambda}{\mu^a_{\delta_{i,j}}} - 1 + \eta_{i,j} \right)^+ = \text{Max}\left(\frac{\lambda}{\mu^a_{\delta_{i,j}}} - 1 + \eta_{i,j}, 0 \right) \quad (4.38)$$

$$\left(\frac{\lambda}{\mu^a_{\sigma_{i,k}}} - 1 + \eta_{i,k+N_d} \right)^+ = \text{Max}\left(\frac{\lambda}{\mu^a_{\sigma_{i,k}}} - 1 + \eta_{i,k+N_d}, 0 \right) \quad (4.39)$$

In the fuzzy augmented Lagrangian GA the constraints are fuzzy with membership functions as in Figure 4.4. The design search is confined to region B in Figures 4.1 and 4.3. The value of the membership function for the objective function (weight of structure) is 1.0 only if it is less than the value of the global optimum to be found, which implies an infeasible design. When the tolerance for the Lagrangian functional is relatively large, say 0.01, the algorithm tends to converge to a feasible solution somewhat larger than the global optimum (minimum weight) solution. If the tolerance value is reduced, the possibility of reaching the global optimum will increase, but at the cost of an increased number of iterations. When the tolerance value is made very small, say 0.0001, some constraints are violated slightly within the fuzzy relaxation limits but the possibility of reaching the global optimum is further increased. However, for very small values of tolerance the solution may require a prohibitively large number of iterations. Thus, selection of an appropriate value for the tolerance is important for the global optimum solution. The efficiency of the algorithm also depends on the selected range for the fuzzy domain, B (Figures 4.1 and 4.3), to be discussed subsequently in this section.

In the present fuzzy GA optimization model λ is considered as a scalar term. To improve the convergence λ is defined as the function of three genetic variables λ_1, λ_2, and λ_3 such that

$$\lambda = \sqrt{(\lambda_1^2 + \lambda_2^2 + \lambda_3^2)/3} \tag{4.40}$$

Like the design variables ($\mathbf{x}$), λ_1, λ_2, and λ_3 are also randomly generated variables, subjected to the genetic operations of crossover, reproduction, and mutation. It was found that this root mean square value of λ improves the convergence for larger structures. The values of λ_1, λ_2, and λ_3 are generated randomly within the range of 0.01 to 1.

The upper and lower limits on the design variables (cross-sectional areas) are the cross-sectional areas of the largest and smallest sections available in the AISC manual (AISC, 1995). These upper and lower bounds are significant in GA-based optimization because the variables are randomly generated within that range. Thus, specifying an appropriate range for these bounds speeds up the convergence for obtaining the global optimum.

An estimate of the range of the global optimum solution (parameter B in Figures 4.1 and 4.3) is needed for the most efficient operation of the fuzzy GA. An estimate of this range and the initial values for the design variables may be based on experience obtained from previous designs. In the absence of such experience the design variables may be generated randomly. The recommended approach, however, is to run the simple augmented Lagrangian GA, discussed in Chapter 2 for a number of iterations, until the convergence curve becomes asymptotic to the iteration axis (Figure 4.5). This solution is then used to choose an appropriate range for the global optimum solution as well as to narrow down the values for the upper and lower bounds on the design variables to be used in the fuzzy GA. This approach helps the search for the global optimum to be done efficiently in a narrow range of sections.

The magnitude of the relaxation or fuzzy range for the upper and lower bounds (Δg_i^u and Δg_i^l) on displacement and stress constraints (equation (4.29)) are chosen as a percentage of the bounds themselves. Selection of appropriate

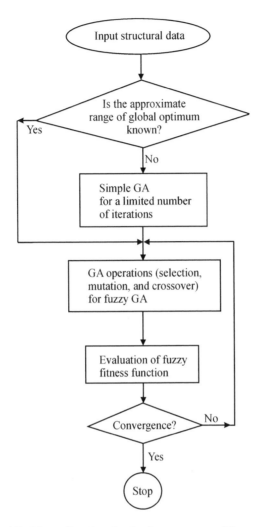

Figure 4.5   Macro flowchart for the fuzzy augmented Lagrangian GA

relaxation ratios (Δ) is important for the convergence of the algorithm. Figure 4.6 demonstrates that with a relatively large value of relaxation or fuzzy range, Δg_i^{u}, a small intrusion into the fuzzy domain, δg_i, yields a high value for the membership function, $\mu_{\delta_{g_i}}$, for the constraint i. Higher values of the membership function ($\mu_{\delta_{g_i}}$) facilitate obtaining the optimum value of λ, as λ should be less than the membership functions (equations (4.6) to (4.8)). In this work relaxation ratios (Δ) are chosen in the range of 0.02 to 0.3. When a relatively large number is chosen for the relaxation

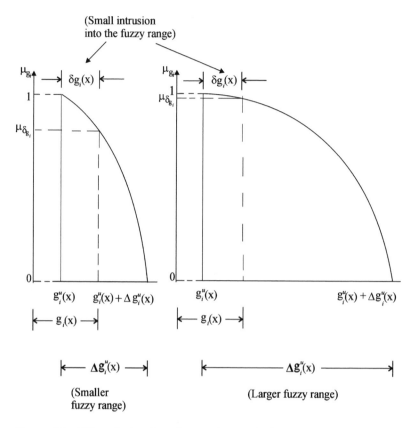

Figure 4.6   Effect of relaxation of constraints on the fuzzy membership functions

ratio, the magnitude of the constraint violation may be relatively large (in the range of 2 % to 3 %), but the convergence takes a smaller number of iterations. To limit the constraint violation to around 1 %, a small value of the relaxation ratio, about 0.02, is used. In this case, the lower bound for the random variables in the functional λ (equation (4.40)) should also be lowered from 0.01 to 0.001 (λ is always less than the values of the membership functions).

4.6 Implementation and Examples

The efficiency of any GA in finding the global optimum also depends on the selection of the crossover operator and crossover and mutation rates. In the examples presented in this chapter the two-point crossover operator is used.

The rate (probability) of crossover is varied from 0.6 to 0.9. The mutation rate is varied between 0.001 and 0.005.

The computational model has been used to find the minimum weight solution for two space axial force structures. The first example, solved on a personal computer, is a small test example with explicit constraints solved by a number of other researchers. The second example is a large steel space structure subjected to the actual constraints of the AISC ASD specifications (AISC, 1995). This example could not be solved on a personal computer within a reasonable amount of time; it was solved on the SGI Origin 2000 supercomputer.

4.6.1 Example 1

This example is a 72-member space truss shown in Figure 4.7. The same problem was studied by Adeli and Kamal (1986), Adeli and Park (1998), and many other researchers. The structure is subjected to two loading conditions. In the first loading condition, node 17 is subjected to 22.24 kN (5.0 kips), 22.24 kN (5.0 kips), and −22.24 kN (−5.0 kips), along the x, y and z directions, respectively. In the second loading condition, nodes 17, 18, 19, and 20 are subjected to −22.24 kN (−5.0 kips) loads along the z axis. The allowable stresses for both tension and compression are 172.4 MPa (25 ksi). The displacement constraints are applied at the four upper nodes 17, 18, 19, and 20 with limiting displacements as ±0.635 cm (0.25 in.) in both the x and y directions. The 72 members of the truss are linked to 16 design variables with a lower bound 0.0645 cm^2 (0.01 in.^{2}) on the cross-sectional area.

The design histories of this example using simple and fuzzy GAs are presented in Figure 4.8. The optimum solution obtained by the simple and the fuzzy GAs are 1.6564 kN (372.40 lb) and 1.6208 kN (364.40 lb) in 2776 and 1758 generations, respectively. The CPU time used for these two methods on a Pentium-Pro processor are 310.05 and 280.14 seconds, respectively. The optimum solutions obtained by the two methods are compared with those of Adeli and Kamal (1986) and Adeli and Park (1998) in Table 4.1. The maximum violations of the constraints in the simple GA and fuzzy GA are 0.004 % and 0.9 %, respectively.

4.6.2 Example 2

This example is a 1310-bar steel space truss shown in Figure 4.9, representing the exterior envelope of a 37-story steel high-rise building structure

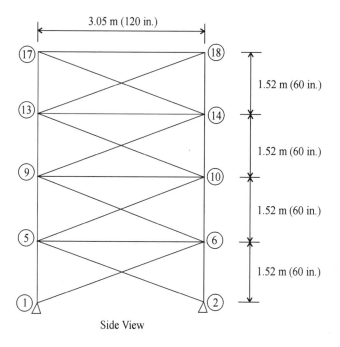

Side View

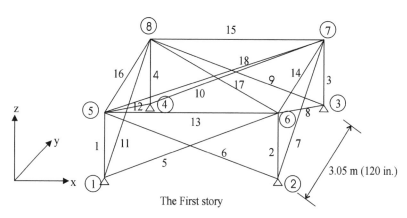

The First story

Figure 4.7   A 72-member space truss

(Adeli and Park, 1995a). It has 332 nodes and 105 groups of members (design variables). In the three main sections of the structure, the same cross-sectional area is used for vertical members of every two stories, the same cross-sectional area is used for horizontal members of equal lengths in each story, and the same cross-sectional area is used for inclined members

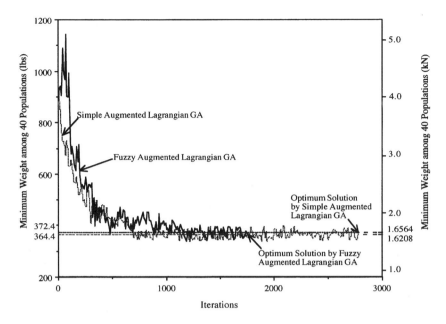

Figure 4.8 Iteration history of the 72-member space truss

in each story. In the two tapered transitional zones there are only inclined members, which are grouped into one variable in each zone. The design is done according to the AISC-ASD specifications (AISC, 1995). A limit on the horizontal drift equal to $\pm 0.5\,\text{m}$ ($\pm 19.69\,\text{in}$) (0.4% of the height of the structure) is used at the topmost nodes along the two horizontal x and y directions. The design constants used are modulus of elasticity, $E = 198.91\,\text{GPa}$ (29 000 ksi), specific weight $\rho = 76.97\,\text{kN/m}^3$ (490.0 lb/ft^3), and yield stress, $F_y = 344.75\,\text{MPa}$ (50 ksi). The loading on the structure consists of both vertical and horizontal loads. A uniform vertical load of $1.92\,\text{kPa}$ (40.0 psf) is used on each floor. The lateral loads due to wind loads are calculated per Uniform Building Code (UBC, 1997). Lateral loads are calculated by the normal force method with basic wind speed 70 mph, exposure C, and importance factor 1. The upper and lower bounds on stresses are obtained from the AISC ASD design specifications (AISC, 1995).

The design histories of this example for simple and fuzzy GAs are presented in Figure 4.10. Adeli and Park (1998) reported a minimum weight of 4130.4 kN (928.6 kips) for this example. The present authors obtained a minimum weight of 4093.0 kN (920.2 kips) using the simple augmented Lagrangian GA. However, using the fuzzy augmented Lagrangian GA

Table 4.1 Optimal solutions in cm^2 (in.^{2}) for the 72-member structure (Example 1)

Member[a]	Adeli and Kamal (1986)	Adeli and Park (1998)	Simple GA $\varepsilon = 0.00001$[b]	Fuzzy GA $\varepsilon = 0.00001$[b]
1, 2, 3, 4	13.0703	17.7722	13.8109	11.1748
	(2.0259)	(2.7547)	(2.1407)	(1.7321)
5, 6, 7, 8, 9, 10,	3.4400	3.2915	3.2890	3.3645
11, 12	(0.5332)	(0.5102)	(0.5098)	(0.5215)
13, 14, 15, 16	0.0645	0.0645	0.3471	0.0645
	(0.0100)	(0.0100)	(0.0538)	(0.0100)
17, 18	0.0645	0.0645	0.0645	0.0832
	(0.0100)	(0.0100)	(0.0100)	(0.0129)
19, 20, 21, 22	7.4626	8.8361	9.6058	8.6780
	(1.1567)	(1.3696)	(1.4889)	(1.3451)
23, 24, 25, 26,	3.6703	3.2708	3.5529	3.5529
27, 28, 29, 30	(0.5689)	(0.5070)	(0.5507)	(0.5507)
31, 32, 33, 34	0.0645	0.0645	0.3665	0.0645
	(0.0100)	(0.0100)	(0.0568)	(0.0100)
35, 36	0.0645	0.0645	0.0832	0.0832
	(0.0100)	(0.0100)	(0.0129)	(0.0129)
37, 38, 39, 40	3.3142	3.1012	3.6471	3.1761
	(0.5137)	(0.4807)	(0.5653)	(0.4923)
41, 42, 43, 44, 45,	3.0910	3.2798	3.4019	3.5155
46, 47, 48	(0.4791)	(0.5084)	(0.5273)	(0.5449)
49, 50, 51, 52	0.0645	0.0645	0.0645	0.4226
	(0.0100)	(0.0100)	(0.0100)	(0.0655)
53, 54	0.0645	0.4148	0.4226	0.0832
	(0.0100)	(0.0643)	(0.0655)	(0.0129)
55, 56, 57, 58	1.0187	1.3875	1.1206	1.1471
	(0.1579)	(0.2151)	(0.1737)	(0.1778)
59, 60, 61, 62,	3.5490	3.3412	2.7419	3.3832
63, 64, 65, 66	(0.5501)	(0.5179)	(0.4250)	(0.5244)
67, 68, 69, 70	2.2252	2.7029	2.8174	2.5535
	(0.3449)	(0.4190)	(0.4367)	(0.3958)
71, 72	3.2155	3.2508	4.1374	3.8400
	(0.4984)	(0.5039)	(0.6413)	(0.5952)
Optimum weight,	1.687	1.675	1.6564	1.6208
kN (lb)	(379.31)	(376.50)	(372.40)	(364.40)
Iterations required			2776	1758

[a] In Figure 4.7 only the members in the first story are shown. The location of any member in the second, third, or fourth stories is obtained by adding 18, 36, or 54, respectively, to the corresponding member number in the first story.
[b] ε = tolerance on the Lagrangian functional.

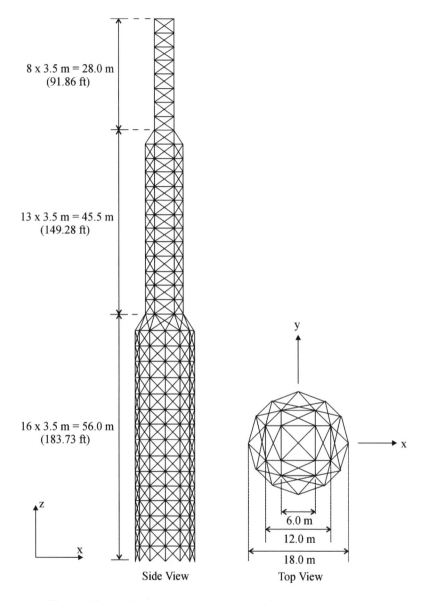

Figure 4.9 A 1310-member steel space truss (Adeli and Park, 1995a)

presented in this work, the minimum weight is reduced further to 4046.3 kN (909.7 kips). It was obtained after 2639 iterations using 400 populations of design variables. The CPU time used in the fuzzy GA is 86 % of that used in the simple GA with 3106 iterations.

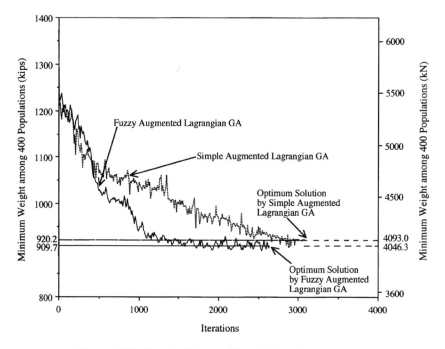

Figure 4.10   Iteration history of the 1310-member space truss

4.7 Conclusion

An augmented Lagrangian fuzzy GA for the minimum weight design of structures was presented in this chapter. The algorithm was applied to two space axial load structures and results were compared with the simple (non-fuzzy) augmented Lagrangian GA. The largest example with 1310 members was subjected to the constraints of the AISC ASD specifications. The maximum constraint violations for the two examples are in the range 0.001 % to 0.004 % for the simple GA and in the range 0.9 % to 1.4 % for the fuzzy GA. The magnitudes of the constraint violations are larger for the fuzzy GA but still small and within practically acceptable limits. Since in practice the continuous optimum design values are rounded up when selecting commercially available discrete sections the actual values of the constraint violation will be further reduced to very negligible values.

The advantages of the fuzzy GA algorithm presented in this chapter can be summarized as follows:

- acknowledgment and incorporation of the imprecision and fuzziness in the code-based design constraints;
- reduction in the value of the optimum weight;
- increasing the likelihood of obtaining the global optimum; and
- reduction in the number of iterations and the total computer processing time needed.

5

Fuzzy Discrete Multi-criteria Cost Optimization of Steel Structures

5.1 Cost of a Steel Structure

The costs of members of a structure are not necessarily proportional to their weights. The unit costs of different structural shapes differ according to the size and the grade of steel and are commonly expressed in the manufacturers' price lists in terms of dollar per 100 lb of weight. Cost data obtained from steel mills indicate that the unit cost usually increases with an increase in the size of a section of a particular grade. However, there are also exceptions; for example, the less frequently used sections may be more expensive than those more frequently used. As an example, one steel company currently prices A36 and A572 Grade 50 W12 $\times$ 65 shapes with cross-sectional area of 19.1 in.^{2}(123.22 cm^2) at \$11.05/ft (\$36.25/m) and the heavier W8 $\times$ 67 shapes with the cross-sectional area of 19.7 in.^{2}(127.10 cm^2) at \$10.05/ft (\$32.97/m) (Nucor, 1999a, 1999b, 1999c).

For many sections made of A36 and A572-50, the two most popular grades of steel in the US, the unit costs of similar sections are the same. However, the unit costs for some sections made of A572-50 are higher than those of grade A36. In general, the heavier and deeper sections cost more. For example, according to one manufacturer (Nucor, 1999a, 1999b, 1999c), sections made of A572-50 with a weight per unit length in the range of 76 lb/ft to 82 lb/ft, in general, cost more by \$1.25 to \$2.25 per 100 lb compared with similar sections

Cost Optimization of Structures: Fuzzy Logic, Genetic Algorithms, and Parallel Computing   H. Adeli and K. C. Sarma   © 2006 John Wiley & Sons, Ltd

made of A36 steel. Steel grade A572-60 is more expensive than A36 in the range of $3.00 to $3.50 per 100 lb. Similarly, steel grade A588 costs more than A36 in the range of $1.25 to $2.50 per 100 lb. Furthermore, not all the sections listed in the AISC manuals (AISC, 1995, 2001) are manufactured in the US. Some uncommon sections are either made on special orders or have to be imported from other countries, adding to their unit costs. Thus, a minimum weight design is not necessarily a minimum cost design even considering the costs of the materials only. Moreover, steel fabricators charge more when a larger number of different types of shapes are used in a given structure.

Cost optimization of realistic steel structures has to be formulated as a discrete optimization problem because these structures consist of discrete commercially available standard shapes. Such a discrete optimization problem may be solved in two stages. First, an approximate continuous optimization solution is found to be followed by the second discrete optimization stage. This is the approach adopted in this book.

In this chapter, a fuzzy discrete multi-criteria optimization algorithm is presented for minimizing the total cost of steel structures by considering three design criteria simultaneously. Factors contributing to the total cost of a steel structure are reviewed in the next section.

5.2 Primary Contributing Factors to the Cost of a Steel Structure

The total cost of a steel structure can be considered as the sum of nine different cost components or functions:

(1) cost of planning and design;
(2) material cost of structural members such as beams, columns, and bracings;
(3) fabrication cost including the material costs of connection elements, bolts, electrodes, and the labor cost;
(4) cost of transporting the rolled sections to the fabrication shop and transporting the fabricated pieces to the construction field;
(5) receiving, handling, and storage costs of rolled sections and fabricated pieces;
(6) erection cost including the material costs of connection elements, bolts, electrodes, and the labor cost;
(7) cost of the operation of tools and machinery on the construction site;
(8) cost of preparing the project site including the cost of the foundation; and
(9) maintenance cost.

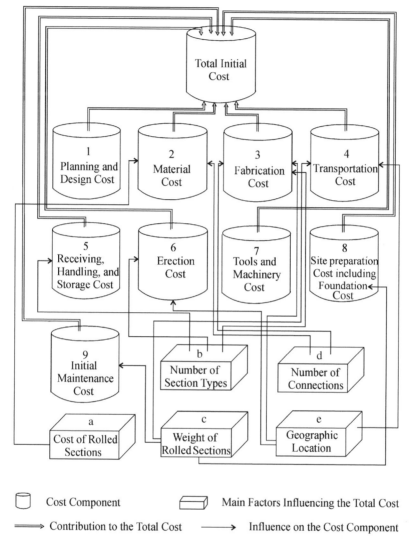

Figure 5.1 Relationship among different cost components and the five influence factors

Figure 5.1 presents a pictorial view of different cost functions and their relationships in contributing to the total cost. Five main factors are identified that influence the total cost of a structure significantly. These are:

(1) cost of the rolled sections;
(2) number of different section types used in the structure;

(3) weight of rolled sections;

(4) number of connections; and

(5) geographic location of the project site.

The influence of these factors on different cost functions is shown in Figure 5.1 with single lines.

Out of the five factors, the cost of rolled sections is often the most important factor in the total cost of a structure. The second factor, the number of different section types, is important in reducing the aforementioned cost functions 3, 5, and 6. For a large structure with hundreds or thousands of members, it costs more to purchase, store, and fabricate relatively small quantities of a large number of different types of rolled sections with small variations in their sizes. A smaller number of rolled section types means a larger purchase order with less paper work and reduced management costs with easier handling and storage. Fabrication work involving labor is much easier with a smaller chance for errors when a smaller number of visibly distinct and different section sizes are used (Templeman, 1988). Further, the amount of scrap sections is reduced. However, reducing the number of section types should not affect the material cost and weight considerably. Where do we draw the line? An attempt is made to formalize an answer to this very significant design question in this chapter.

The third factor, the weight of rolled sections, is also significant for reducing the total overall cost of a structure as it influences the transportation cost, the cost of the substructure, and to a smaller extent the maintenance cost. The transportation charges depend primarily on the weight and the distance traveled. Similarly, the cost of the foundation depends on the weight of the superstructure.

The fourth factor, the number of connections, influences the costs of fabrication, erection, connection materials, and labor. The number and location of connections are decided by the architect and engineer at the planning stage and configuration design of the structure. Once the configuration has been selected the designer has little control over the number of connections. However, he can decide to limit the number of different types of connections. This can reduce the fabrication cost. The number of different connection types may also be reduced when the number of member section types is reduced.

The fifth factor, the geographic location of the site, influences the aforementioned cost functions 4, 5, 6, 7, and 8. A place with an abundance of a skilled and unskilled labor force will cost less than a place where the labor force is scarce and expensive. Geographic location also influences the transportation cost of rolled sections from the steel mills to the fabrication shop and subsequently to the construction sites. In urban areas, the truck traffic may be limited by size and time of the day (Means, 1999). The structural designer usually has no control over this factor.

In this chapter, the first three factors are considered and a three-criteria cost optimization model is formulated for the design of steel structures. These criteria are (a) select discrete commercially available sections with the lowest cost, (b) select discrete commercially available sections with the lightest weight, and (c) select a minimum number of different types of discrete commercially available sections.

In the cost optimization of structures additional difficulties are encountered while defining the cost function due to imprecision, fuzziness, and uncertainties involved in determining the cost parameters. Therefore, the formulation of the cost function and its subsequent solution becomes complicated. This complexity is overcome by using the fuzzy logic (Zadeh, 1965). To apply the fuzzy logic in the cost optimization of steel structures the design variables, the cross-sectional areas, are considered as fuzzy. In the next section a methodology and computational model are presented for fuzzy discrete multi-criteria cost optimization of steel structures using the aforementioned three criteria.

5.3 Fuzzy Discrete Multi-criteria Cost Optimization

The fuzzy discrete multi-criteria cost optimization model consists of two stages. In the first stage, an initial minimum weight solution, $\mathbf{x}$, is found using the fuzzy augmented Lagrangian genetic algorithm (GA) developed in Chapter 4 assuming continuous variables. Each continuous cross-sectional area obtained from this stage forms the basis for selecting a few candidate commercially available discrete sections with fuzzy attributes or membership functions for each of the aforementioned three criteria. Using the max-min procedure of Bellman and Zadeh (1970) on the three fuzzy criteria, the best discrete section for each design variable is selected.

The second criterion (minimum number of section types) usually conflicts with the first and third criteria (minimum material cost and minimum weight). Minimum weight and minimum material cost designs are not always the same, as explained in the introduction. As such, the first and third criteria may be conforming or conflicting. This methodology handles the relative importance of the three different criteria through introduction of a weight coefficient for each criterion with a total sum of one.

Three fuzzy functions are defined: the material cost of the structure $\tilde{C}(\tilde{y})$, the weight of the structure $\tilde{W}(\tilde{y})$, and the number of different section types

$\tilde{T}(\tilde{y})$ in terms of the fuzzy discrete variables (commercially available discrete shapes), $\tilde{y}$. These are fuzzy functions (identified by the wavy sign $\sim$ on the top) because the variables $\tilde{y}$ are treated as fuzzy variables. The objective of the three-criteria optimization is to minimize the functions $\tilde{C}(\tilde{y})$, $\tilde{W}(\tilde{y})$, and $\tilde{T}(\tilde{y})$. Out of these three fuzzy functions $\tilde{C}(\tilde{y})$ and $\tilde{W}(\tilde{y})$ can be expressed explicitly in terms of the fuzzy variables $\tilde{y}$, and are presented below:

$$\tilde{C}(\tilde{y}) = \sum_{i=1}^{N_t} l_i c_i \tilde{y}_{C_i}, \qquad \tilde{y}_{C_i} \in S_{C_i} \tag{5.1}$$

where S_{C_i} is a fuzzy set of discrete candidate standard shapes for the design variable x_i corresponding to the minimum cost criterion. In equation (5.1), N_t is the number of initial section types (equal to the number of design variables in the first continuous variable stage of the optimization), l_i is the total length of members linked to the variable x_i, $\tilde{y}_{C_i}$ is the cross-sectional area of the discrete standard shape with the maximum membership function corresponding to the minimum cost criterion and belonging to the fuzzy set S_{C_i}, and c_i is the cost per unit volume of the standard shape $\tilde{y}_{C_i}$. Similarly,

$$\tilde{W}(\tilde{y}) = \rho \sum_{i=1}^{N_t} l_i \tilde{y}_{W_i}, \qquad \tilde{y}_{W_i} \in S_{W_i} \tag{5.2}$$

where ρ is the specific weight of steel, $\tilde{y}_{W_i}$ is the cross-sectional area of the discrete standard shape with the maximum membership function corresponding to the minimum weight criterion, and S_{W_i} is a set of fuzzy discrete candidate standard shapes for the design variable x_i corresponding to the minimum weight criterion. The numbers of candidate standard shapes in the set S_{C_i} or S_{W_i} may be different for different design variables and are chosen based on certain criteria to be discussed in the following section.

In the first stage of optimization when the design variables, **x**, are continuous the number of section types is constant. In the second discrete multi-criteria optimization stage the number of section types, $\tilde{T}(\tilde{y})$, is treated as a fuzzy variable with an upper limit of N_t. The goal is possibly to minimize this value along with the material cost and the weight of the structure.

The three-criteria cost optimization model is formulated as: find the vector of discrete cross-sectional areas of members, $\tilde{y}|\tilde{y} \in S$ where S is the set of all commercially available sections, to minimize the three fuzzy objective functions $\tilde{C}(\tilde{y})$, $\tilde{W}(\tilde{y})$, and $\tilde{T}(\tilde{y})$ subject to the displacement constraint:

$$\delta_i \leq \delta_i^{\mathrm{a}} \quad \text{for } i = 1, 2, \ldots, N_{\mathrm{d}} \tag{5.3}$$

and the stress constraints according to the AISC ASD code (AISC, 1995). In the most general case of three-dimensional moment-resisting frames, the stress constraints are in the form of the following highly nonlinear and discontinuous interaction equations:

$$\frac{f_{a_i}}{F_{ac_i}} + \frac{C_{mx} f_{bx_i}}{(1 - f_{a_i}/F'_{ex})F_{bx}} + \frac{C_{my} f_{by_i}}{(1 - f_{a_i}/F'_{ey})F_{by}} \le 1.0$$

$$\frac{f_{a_i}}{0.6F_y} + \frac{f_{bx_i}}{F_{bx}} + \frac{f_{by_i}}{F_{by}} \le 1.0$$

$$\text{for } f_{a_i} < 0.0 \text{ and } \frac{f_{a_i}}{F_{ac_i}} > 0.15 \qquad (5.4)$$

$$\frac{f_{a_i}}{F_{ac_i}} + \frac{f_{bx_i}}{F_{bx_i}} + \frac{f_{by_i}}{F_{by_i}} \le 1.0 \quad \text{for } f_{a_i} < 0.0 \text{ and } \frac{f_{a_i}}{F_{ac_i}} \le 0.15 \qquad (5.5)$$

where δ_i is the displacement of the ith nodal displacement degree of freedom and the allowable displacement δ_i^a is expressed by equations (2.16a) and (2.16b). In equations (5.4) and (5.5), f_{a_i}, f_{bx_i}, and f_{by_i} are the computed axial stress and the bending stresses about the x and y axes of the ith member. Similarly, F_{ac_i}, F_{bx_i}, and F_{by_i} are the allowable axial compressive stress and the allowable bending stresses about the x and y axes, respectively. The horizontal girder members are assumed to have full lateral supports. However, for columns and bracings, lateral supports are provided at the ends only. For computation of the allowable axial stress and the terms

$$F'_{ex} = \frac{12\pi^2 E}{23 (Kl_b/r_{bx})^2} \qquad \text{and} \qquad F'_{ey} = \frac{12\pi^2 E}{23 (Kl_b/r_{by})^2}$$

the effective length factor K is needed. The AISC ASD and LRFD codes (AISC, 1995, 2001) provide alignment charts for finding the values of K based on the numerical solution of transcendental equations. In this book, the values of K are found by the following approximate but explicit equations used in the European steel design code (Anonymous, 1978; Dumonteil, 1992). For braced frames,

$$K = \frac{3G_A G_B + 1.4(G_A + G_B) + 0.64}{3G_A G_B + 2.0(G_A + G_B) + 1.28} \qquad (5.6)$$

and for unbraced frames,

$$K = \sqrt{\frac{1.6G_A G_B + 4.0(G_A + G_B) + 7.50}{G_A + G_B + 7.50}} \qquad (5.7)$$

where subscripts A and B refer to the two ends of a member and factor G is defined as

$$G = \frac{\sum I_c/L_c}{\sum I_g/L_g} \tag{5.8}$$

in which I_c, I_g, L_c, and L_g are the moments of inertia and unsupported lengths of columns and beams, respectively, and Σ represents summation of all members connected to a joint and lying in the plane of buckling.

In the fuzzy discrete multi-criteria cost optimization model, for each continuous variable $x_i (i = 1, 2, \ldots, N_t)$, N_i commercially available sections, $\tilde{y}_{ij} \geq x_i (j = 1, 2, \ldots, N_i)$ are chosen. All the N_i shapes selected are graded with membership functions for each of the three fuzzy criteria. The expressions for the membership functions will be presented in the next section. Thus, S_{C_i}, S_{W_i}, and S_{T_i} are the three fuzzy sets of candidate discrete shapes corresponding to the three aforementioned criteria for the design variable x_i. To find the multi-criteria optimum solution, the max-min procedure of Bellman and Zadeh (1970) is used, described in Section 4.2. The solution is found by forming a fuzzy set S_{D_i} defined as the intersection of the three fuzzy sets S_{C_i}, S_{W_i}, and S_{T_i} as follows:

$$S_{D_j}(\tilde{y}_{ij}) = S_{C_j}(\tilde{y}_{ij}) \cap S_{W_j}(\tilde{y}_{ij}) \cap S_{T_j}(\tilde{y}_{ij}), \quad j = 1, 2, \ldots, N_i \tag{5.9}$$

The membership of this intersection fuzzy set S_{D_i} is set equal to the minimum values of the membership functions of the three sets S_{C_i}, S_{W_i}, and S_{T_i}:

$$\mu_{D_j}(\tilde{y}_{ij}) = \min\left[\mu_C(\tilde{y}_{ij}); \mu_W(\tilde{y}_{ij}); \mu_T(\tilde{y}_{ij})\right], \quad j = 1, 2, \ldots, N_i \tag{5.10}$$

where $\mu_C(\tilde{y}_{ij})$, $\mu_W(\tilde{y}_{ij})$, and $\mu_T(\tilde{y}_{ij})$ are the fuzzy membership functions for the three criteria: minimizing material cost, minimizing weight, and minimizing the number of section types, respectively. The commercially available discrete section corresponding to the maximum membership function, $\mu_{D_{i(best)}}$, in the N_i selected candidate shapes represents the best section (Klir and Folger, 1988):

$$\mu_{D_{i(best)}} = \max_{j=1,2,\ldots,N_i}\left[\mu_{D_j}(\tilde{y}_{ij})\right], \quad i = 1, 2, \ldots, N_t \tag{5.11}$$

This step is known as defuzzification. This concept can be further explained through an example, presented in Figure 5.2. In this figure, six candidate commercially available shapes are considered for a particular design variable (y_1, y_2, y_3, y_4, y_5, and y_6). For each variable $y_i (i = 1, 2, \ldots, 6)$,

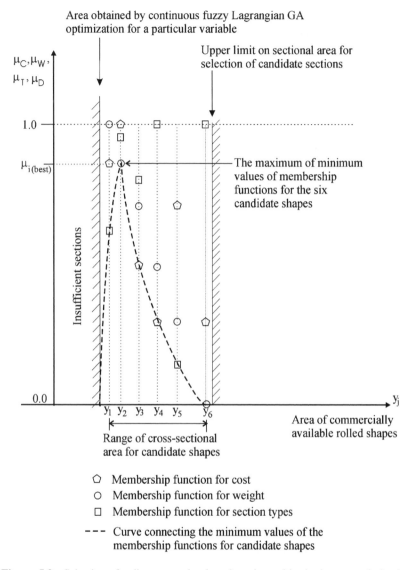

Figure 5.2 Selection of a discrete section based on the multi-criteria cost optimization procedure

three different values of membership functions (μ_C, μ_W, and μ_T) corresponding to three different criteria are computed. The dashed curve shown in Figure 5.2 connecting the minimum values of the membership functions represents the membership function for the intersection fuzzy domain, μ_D.

The peak of this curve represents $\mu_{D_{i(best)}}$. The commercially available section corresponding to this value is selected as the best shape among the candidate shapes. Thus, by using the max-min operation of fuzzy membership functions (equations (5.10) and (5.11)), the best commercially available discrete shapes are obtained for all the design variables. In a subsequent section it is proved mathematically that this solution is Pareto optimal.

5.4 Membership Functions

5.4.1 Membership Function for Minimum Cost

Denoting the material cost per unit length for the jth candidate shape by c_j and the maximum and the minimum values within the N_i candidate shapes by $c_{j_{max}}$ and $c_{j_{min}}$, a linear membership function is defined in the following form (Figure 5.3a):

$$\mu_{C_j} = 1 - \frac{c_j - c_{j_{min}}}{c_{j_{max}} - c_{j_{min}}}, \quad j = 1, 2, \ldots, N_i \tag{5.12}$$

5.4.2 Membership Function for Minimum Weight

Denoting the cross-sectional area of the jth candidate section by y_j and the maximum and the minimum values within the N_i candidate shapes by $y_{j_{max}}$ and $y_{j_{min}}$, a linear membership function is defined in the following form (Figure 5.3b):

$$\mu_{W_j} = 1 - \frac{y_j - y_{j_{min}}}{y_{j_{max}} - y_{j_{min}}}, \quad j = 1, 2, \ldots, N_i \tag{5.13}$$

5.4.3 Membership Function for Minimum Number of Section Types

For the design of multi-story high-rise building structures the individual members are designed from the bottom to the top. The goal is to minimize the number of different section types without adversely affecting the total material cost of the structure. Records of the selected standard shapes are kept so that the same sectional shape may be selected again for the design of subsequent members. The following scheme is used for assigning the

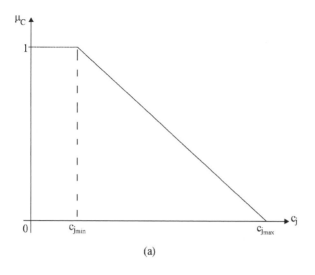

(a)

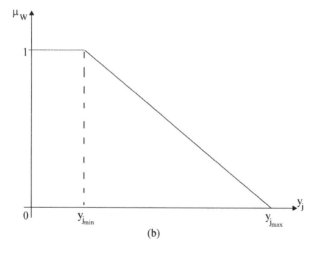

(b)

Figure 5.3 Membership function for (a) minimum material cost and (b) minimum weight

membership values to the candidate sections for minimizing the number of section types:

$$
\mu_{T_j} = \begin{cases} \beta\mu_{C_j} & \text{if the section is never selected} \\ \gamma\beta\mu_{C_j} & \text{if the section is selected earlier and } \beta\mu_{C_j} \leq 1.0/\gamma \\ 1.0 & \text{if the section is selected earlier and } \beta\mu_{C_j} > 1.0/\gamma \end{cases}
$$

(5.14)

where $\gamma = 1 + e^{n_f}/\alpha$ and n_f is the number of times a particular shape is used earlier, β is a factor used to penalize a shape if it is not used before, and α is a scaling factor. It was found that $\beta = 0.2$ and $\alpha = 10$ to be the most suitable values based on parametric studies. The penalizing and the scaling factors, β and α, are chosen so that the value of the exponential expression for γ helps in choosing a candidate shape only if the same shape has been used for some other members at least four times.

5.5 Fuzzy Membership Functions for Criteria with Unequal Importance

In obtaining the intersection fuzzy set $S_{D_j}(\tilde{y}_{ij})$ in equation (5.9), it was assumed that all the three criteria were of equal importance. However, some of them may be of greater importance than others. In such cases, $S_{D_j}(\tilde{y}_{ij})$ may be expressed as a convex combination of the three criteria with weighting coefficients reflecting their relative importance. In this book three weighting coefficients w_C, w_W, and w_T are used for minimizing the material cost, sectional weight, and the number of section types, respectively. It should be noted that since the first step of the fuzzy max-min procedure is selecting the minimum values of the membership functions, there is an inverse relationship between the weighting coefficients and their impact on the multi-criteria cost optimization. In other words, to increase the importance of a criterion its weighting coefficient is reduced.

The modified membership functions are defined as

$$
\begin{aligned}
\mu'_{C_i} &= w_C \mu_{C_i} \\
\mu'_{W_i} &= w_W \mu_{W_i} \\
\mu'_{T_i} &= w_T \mu_{T_i}
\end{aligned}
\tag{5.15}
$$

where the sum of the weighting coefficients is equal to 1:

$$
w_C + w_W + w_T = 1
\tag{5.16}
$$

5.6 Pareto Optimality

In general a multi-criteria optimization problem is defined as

$$
\min\{F_1(\mathbf{y}), F_2(\mathbf{y}), F_3(\mathbf{y}), \ldots, F_k(\mathbf{y})\}
\tag{5.17}
$$

where $F_i(\mathbf{y})(i = 1, 2, \ldots, k)$ is the ith objective functions in terms of variables $\mathbf{y} \in \mathbf{Y}$, $\mathbf{Y}$ is the set of feasible solutions in the design space R^n, and k is the number of objective functions. The fundamental problem is that these objective functions may be contradictory. The concept of the Pareto optimum solution is used to find a solution for a multi-criteria optimization problem (Adeli, 1994).

A vector $\mathbf{y}^* \in \mathbf{Y}$ is Pareto optimal if and only if there exists no other feasible solution $\mathbf{y} \in \mathbf{Y}$ such that $F_i(\mathbf{y}) \leq F_i(\mathbf{y}^*)$ for $i = 1, 2, \ldots, k$, and $F_j(y) < F_j(y^*)$ for at least one value of j (Koski, 1994). Such a Pareto optimal solution means that no other feasible vector $\mathbf{y}$ exists that can decrease some criterion without increasing simultaneously another criterion. Suppose in equation (5.17) that $\mathbf{Y}$ is a set of fuzzy variables. The Pareto optimal solution $\mathbf{y}^* \in \mathbf{Y}$ for the fuzzy multicriteria optimization problem is defined as: no other feasible solution exists for $\mathbf{y}$ where $\mathbf{y} \in \mathbf{Y}$ such that $\mu_{F_i}(\mathbf{y}) \geq \mu_{F_i}(\mathbf{y}^*)$ for $i = 1, 2, \ldots, k$ and $\mu_{F_j}(y) > \mu_{F_j}(y^*)$ for at least one other value of j (Dhingra *et al.*, 1992). In other words, for a fuzzy multi-criteria optimization problem, the Pareto optimal $\mathbf{y}^*$ can be stated as: no other feasible solution $\mathbf{y}$ exists whose membership function will increase in some criterion without simultaneously decreasing the membership function for another criterion.

Different approaches have been proposed for generating Pareto optimal solutions such as the sequential optimization method, the linear weighting method, the minimax method, the constrained method, compromise programming, goal programming, and the multi-attribute utility method (Jendo, 1990; Koski, 1994). It is shown that the max-min fuzzy multi-criteria optimization approach used in this book is equivalent to the minimax method used in generating the Pareto optimal solutions.

The general expression for Pareto optimum solutions in a minimax method is (Koski, 1994)

$$F_i(\mathbf{y}^*) = \min \max \left[w_i \hat{F}_i(y) \right], \quad i = 1, 2, \ldots, k \tag{5.18}$$

where

$$\hat{F}_i(y) = \frac{F_i(y) - \min F_i(y)}{\max F_i(y) - \min F_i(y)} \tag{5.19}$$

and w_i is the weight parameter, such that $\sum_{i=1}^{k} w_i = 1$. The values of $\hat{F}_i(y)$ lie between 0 and 1, similar to the values of the membership functions in the fuzzy multi-criteria optimization method. The primary difference is that in the minimax method the goal is to obtain a value of $\hat{F}_i(y)$ as close to 0 as possible, whereas in the fuzzy multi-criteria optimization the goal is to obtain a value of $\mu_{F_i}(\mathbf{y}^*)$ as close to 1 as possible. Using the fuzzy Pareto concept equations (5.18) and (5.19) can be modified as

$$\mu_{F_i}(\mathbf{y}^*) = \min \max \left(w_i \, \mu_{\hat{F}_i(y)} \right), \quad i = 1, 2, \ldots, k \tag{5.20}$$

where

$$\mu_{\hat{F}_i(y)} = \frac{\mu_{F_i(y)} - \max \mu_{F_i(y)}}{\min \mu_{F_i(y)} - \max \mu_{F_i(y)}} = \frac{\max \mu_{F_i(y)} - \mu_{F_i(y)}}{\max \mu_{F_i(y)} - \min \mu_{F_i(y)}} \tag{5.21}$$

In equation (5.21) $\max \mu_{F_i(y)} = 1$ and $\min \mu_{F_i(y)} = 0$. Therefore, equation (5.20) can be written as

$$\mu_{F_i}(\mathbf{y}^*) = \min \max \left[w_i (1.0 - \mu_{\hat{F}_i(y)}) \right]$$
$$= \max \min \left(w_i \, \mu_{\hat{F}_i(y)} \right), \quad i = 1, 2, \ldots, k \tag{5.22}$$

The present fuzzy solution of equations (5.10), (5.11), and (5.15) bear a complete analogy with the Pareto solution of equation (5.22). Therefore, it is concluded that the multi-criteria optimum solution obtained by equation (5.11) is Pareto optimum.

5.7 Selection of Commercially Available Discrete Shapes

Figure 5.4 shows a macro flowchart of the fuzzy discrete multi-criteria optimization model. At the beginning a continuous variable minimum weight solution is obtained using the fuzzy augmented GA of Chapter 4. The solution obtained from this stage can violate the constraints by a predetermined maximum value, say in the range of 2 % to 3 %, because it is used only as a preliminary design. Thus, in the fuzzy discrete multi-criteria optimization model a strategy of preemptive constraint violation is adopted. It is called preemptive because the constraints are violated within a small range in the initial design stage before the final discrete optimization is done. The goal here is to enhance the chance of finding the global discrete optimum and at the same time reduce the computational processing time.

In the conventional crisp optimization the continuous optimum solutions are rounded up or down to obtain the discrete design solutions. Next, the constraints are re-evaluated for any possible violation. If any constraint is violated the discrete design is altered. A branch-and-bound operation is adopted with discrete design enumeration, which requires a lot of redesigns involving structural analyses and computer processing time. On the other

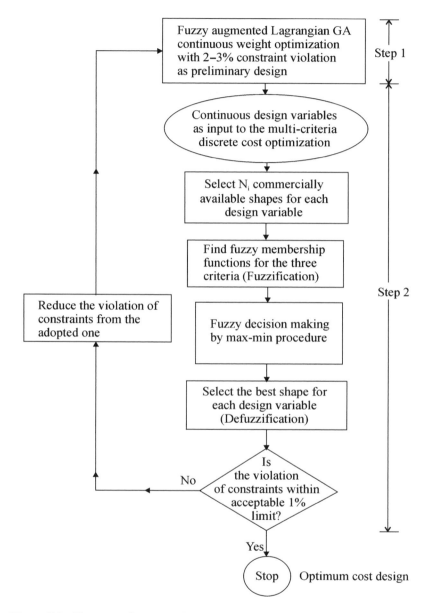

Figure 5.4 The macro flowchart of the fuzzy discrete multi-criteria cost optimization model

hand, if the continuous variable solutions are all rounded up the design will
be safe, but most probably not the optimum or most economical solution.
The preemptive constraint violation strategy is devised to greatly enhance
the possibility of obtaining the global minimum cost solution. This point is
illustrated in Figure 5.5 for a two-variable single criterion constrained optimization problem. It must be pointed out that this strategy reduces the number
of constraints re-evaluation requiring time-consuming structural re-analyses.
The result is a more efficient algorithm.

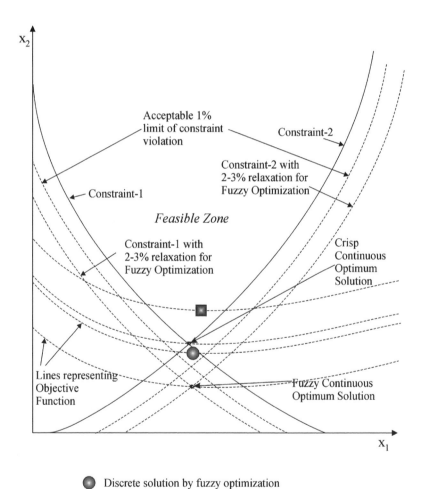

● Discrete solution by fuzzy optimization

■ Discrete solution by crisp optimization

Figure 5.5 Preemptive constraint violation strategy

The values obtained from the continuous variable minimum weight design are used as lower bound solutions in the second discrete multi-criteria optimization stage (Figure 5.2). Then, at least five or six commercially available candidate shapes are selected just above this lower bound from the AISC manual (AISC, 1995). If only one standard shape is selected for each design variable the design becomes the minimum weight design. The same minimum weight design is obtained by setting the weighting parameter w_W to a minimum value, say 0.01, even when the number of candidate shapes (N_i) is more than one. Similarly, by setting the weighting parameter w_C to a minimum value (0.01) the minimum cost design is obtained. A minimum number of five or six potential candidates (N_i) is necessary for the effectiveness of the minimum material cost criterion. On the other extreme, when the number of candidate shapes (N_i) is increased, the minimum section-type criterion would dominate the design process at the expense of the other two criteria.

Increasing the value of N_i reduces the number of section types. It was found that a value of 15 for N_i is a good number for reducing the number of section types in a large structure like the 36-story structure of Example 2, presented subsequently in this chapter. In the actual design of structures the designer should select N_i in consultation with the steel fabricator as the fabricator can determine its impact on reducing the fabrication and handling costs. A higher number of N_i helps in choosing similar shapes used for other members because a wider range of candidate shapes facilitates the selection of already used shapes.

It should be pointed out that in the second stage of the fuzzy multi-criteria discrete optimization model all the constraints are satisfied within a practically acceptable tolerance, say 1 %.

5.8 Implementation and a Parametric Study

The fuzzy multi-criteria discrete cost optimization model presented in this chapter has been implemented in the C language using the IRIX operating system on the SGI Origin 2000 supercomputer at the Ohio Supercomputer Center.

A decision has to be made on the relative values of the importance weighting coefficient for the three membership functions. A parametric study was performed using the following four cases:

Case A $w_C = 0.01$; $w_W = 0.495$; $w_T = 0.495$ ($N_i = 5$ to 7 for Example 1 and 6, 10, and 15 for Example 2; minimum material cost design)

Case B $w_C = 0.495$; $w_W = 0.01$; $w_T = 0.495$ ($N_i = 5$ to 7 for Example
            1 and 6, 10, and 15 for Example 2; minimum weight design)

Case C $w_C = 0.495$; $w_W = 0.495$; $w_T = 0.01$ ($N_i \geq 5$; strong preference
            for the minimum section type design)

Case D $w_C = 0.333$; $w_W = 0.333$; $w_T = 0.333$ ($N_i \geq 5$; equal preference
            for the three criteria)

For columns and axially loaded members, A572 Grade 50: W6, W8, W10, W12, and W14 shapes are used. Similarly, for beams, W16, W18, W21, W24, W27, W30, W33, W36, and W40 shapes from the same grade of steel are used. The design constants used are modulus of elasticity $E = 198.91$ GPa (29 000 ksi), specific weight $\rho = 76.97$ kN/m^3 (490.0 lb/ft^3), and yield stress $F_y = 344.75$ MPa (50 ksi).

5.9 Application to High-Rise Steel Structures

The computational model presented in this chapter has been applied to find the multi-criteria optimization solution for two steel structures, one axially loaded space truss structure and another space moment-resisting frame with bracings. The basis of design is the AISC ASD code (AISC, 1995). The material cost values based on data obtained from Nucor (1999a, 1999b, 1999c) are summarized in Appendix A for axial load members (in Example 1) and columns (in Example 2) and in Appendix B for beams (in Example 2).

5.9.1 Example 1

This example is the same 1310-bar steel space truss shown in Figure 4.9, representing the exterior envelope of a 37-story steel high-rise building structure. The optimum design solutions obtained by the fuzzy multi-criteria discrete optimization model for cases A, B, C, and D are presented in Table 5.1. The minimum weight solutions reported previously by Adeli and Park (1998), and presented in Chapter 4, are also given in the same table. The Adeli and Park solution is a discrete minimum weight solution. The solutions in Chapter 4 are continuous variable solutions and are rounded up to the nearest discrete solutions for the sake of comparison with the current study. Case B of the current study yields the minimum weight solution. The minimum material cost solution, case A of the current study, is about 1.7 % less expensive than the minimum weight solution of Adeli and Park (1998).

Table 5.1 Optimum design solutions obtained by the fuzzy discrete multi-criteria cost optimization model based on the ASD code for the 1310-member structure (Example 1)

Type of study	Material cost of the structure	Weight of the structure	Total number of section types
Case A for $N_i = 5, 6, 7$	$1 678 424	4093.0 kN (920.2 kips)	17
Case B for $N_i = 5, 6, 7$	$1 688 853	4083.7 kN (918.1 kips)	20
Case C(i) for $N_i = 5$	$1 692 032	4113.1 kN (924.7 kips)	11
Case C(ii) for $N_i = 6$	$1 700 052	4131.3 kN (928.8 kips)	10
Case C(iii) for $N_i = 7$	$1 706 295	4146.9 kN (932.3 kips)	8
Case D(i) for $N_i = 5$	$1 691 631	4111.7 kN (924.4 kips)	13
Case D(ii) for $N_i = 6$	$1 693 673	4116.2 kN (925.4 kips)	12
Case D(iii) for $N_i = 7$	$1 694 711	4118.4 kN (925.9 kips)	11
Continuous minimum weight design from Chapter 4 (theoretical optimum weight = 4046.3 kN (909.7 kips))	$1 694 559[a]	4096.2 kN[a] (920.9 kips)	22
Adeli and Park (1998) (discrete minimum weight design)	$1 707 964	4130.4 kN (928.6 kips)	14

[a] For discrete sections obtained by rounding up the minimum weight optimum sections.

The minimum material cost solution for this example is not substantially less expensive than the minimum weight solution. This can be explained partly by the fact that, in this particular space truss example, the number of section types used in the minimum cost design is close to the number of section types used in the minimum weight design. Also, the section database included a relatively small number of wide flange shapes.

Cases C and D of the current study present the solution for a minimum section type and equal importance of the three criteria. The increases of the material costs with the reduction in number of the section types by increasing the value of N_i are studied for these two cases and are noted in Table 5.1. The optimum solution obtained for case C(i) with only 11 different types of shapes is about 1 % less expensive than the minimum weight solution of Adeli and Park (1998) with 14 different types of shapes.

5.9.2 Example 2

This example (Figures 5.6 and 5.7) is a 36-story irregular moment-resisting steel space frame structure with setbacks and cross-bracings and an aspect

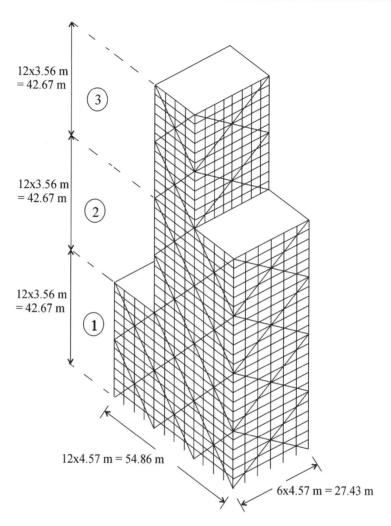

12x3.56 m = 42.67 m ③

12x3.56 m = 42.67 m ②

12x3.56 m = 42.67 m ①

12x4.57 m = 54.86 m

6x4.57 m = 27.43 m

Figure 5.6 A 36-story steel space moment-resisting frame structure (Park and Adeli, 1997a)

ratio of 4.7. A minimum weight solution for the same example was first presented by Adeli and Park (1998). It has 1384 nodes and 3228 members, which are linked to 186 groups of initial design variables. The structure consists of three 12-story sections. In the lower sections 1 and 2, there are four groups of columns (Figure 5.7): corner columns, outer columns, inner columns in the unbraced frames, and inner columns in the braced frames. In section 3, only the first three types of columns are used. The beams in every floor are grouped separately. In sections 1 and 2 they are divided into three groups: outer

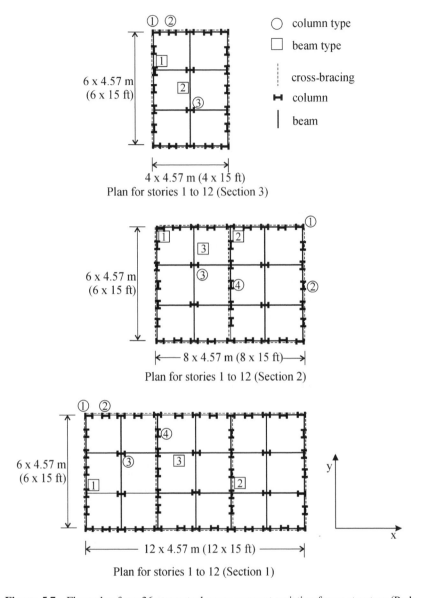

Figure 5.7 Floor plan for a 36-story steel space moment-resisting frame structure (Park and Adeli, 1997a)

beams, inner beams in braced frames, and inner beams in unbraced frames. The beams in section 3 are divided into two groups: inner and outer beams. In every three stories two different bracings are used with the same cross-section, one in the longitudinal direction and the other in the transverse direction.

The inter-story drift is limited to 0.004 times the story height along the two horizontal directions. The dead load and the live load intensities on each floor are 2.88 kPa (60 psf) and 2.38 kPa (50 psf), respectively. The lateral loads due to wind are calculated according to the Uniform Building Code (UBC, 1997), with a basic wind speed of 113 km/h (70 mph), exposure C, and an importance factor of 1.

The optimum design solutions obtained by the fuzzy multi-criteria discrete optimization model for cases A, B, C, and D are presented in Table 5.2. The minimum weight solution reported previously by Adeli and Park (1998) is also given in the same table. Like Example 1, the Adeli and Park solution is a discrete minimum weight solution. Case B of the current study yields the minimum weight solution. The minimum material cost solution, case A of the current study, is 28.8 % less expensive than the minimum weight solution of Adeli and Park (1998), but the number of different types of shapes is 50 compared with 21 in the Adeli and Park study.

Cases C and D of the current study present the solution for the minimum section type and equal importance of the three criteria. The increases of the material costs with the reduction in number of the section types by increasing the value of N_i are also studied for these two cases and are noted in Table 5.2. The optimum solution obtained for case C(iii) with only 19 different types of shapes is about 15.6 % less expensive than the minimum weight solution of Adeli and Park (1998) with 21 different types of shapes.

Table 5.2 Optimum design solutions obtained by the fuzzy discrete multi-criteria cost optimization model based on the ASD code for a 36-story structure (Example 2)

Type of study	Material cost of the structure	Weight of the structure	Total number of section types
Case A for $N_i = 6$, 10, 15	$6 543 527	15516.0 kN (3488.3 kips)	50
Case B for $N_i = 6$, 10, 15	$6 733 198	15410.1 kN (3464.5 kips)	58
Case C(i) for $N_i = 6$	$6 715 814	15938.1 kN (3583.2 kips)	41
Case C(ii) for $N_i = 10$	$7 174 025	16915.7 kN (3803.0 kips)	26
Case C(iii) for $N_i = 15$	$7 758 512	18160.7 kN (4082.9 kips)	19
Case D(i) for $N_i = 6$	$6 685 370	15869.6 kN (3567.8 kips)	41
Case D(ii) for $N_i = 10$	$6 968 327	16511.0 kN (3712.0 kips)	27
Case D(iii) for $N_i = 15$	$7 339 931	17229.8 kN (3873.6 kips)	23
Adeli and Park (1998) (discrete minimum weight design)	$9 196 083	21513.2 kN (4836.6 kips)	21

5.10 Concluding Comments

The goal of this work has been to advance the field of structural optimization for the optimum design of large structures under practical and realistic conditions. For years steel fabricators have pointed out that a minimum weight design is often not the most economical design. In particular, they have pointed out that limiting the variety of shapes used in a structure can result in substantial savings in the fabrication cost. They have, in fact, charged less when the designer used a fewer number of section types or components. A computational model has been presented for the multi-criteria cost optimization of structures considering three different criteria simultaneously: minimum cost, minimum weight, and minimum number of section types. While the computational model is based on advanced computational technologies including fuzzy logic, genetic algorithm, and multi-criteria optimization, it is used to solve a real-world problem of great interest to design engineers.

An important conclusion of this study is that solving the structural design problem as a cost optimization problem can result in substantial cost savings compared with the traditional weight optimization solution, especially for large moment-resisting structures with hundreds or thousands of members. In the second example, a 36-story moment-resisting frame with bracings, the cost savings are substantial and in the range of 15.6 % to 28.8 %.

6

Parallel Computing

6.1 Multiprocessor Computing Environment

The high-performance computing environment used in this book for optimization of very large building structures is the Origin 2000 multiprocessor, a distributed shared memory multiprocessor machine with a hypercube architecture in terms of routers (Figure 6.1), developed by Silicon Graphics Inc. (SGI, 2000a, 2000b, 2000c). A router (identified by R in Figure 6.1) is a switch that allows simultaneous multiple transactions among the processors. Each node board, identified by N in Figure 6.1, is a cluster of two processors, each with its own cache, but with a common main memory and an associated directory used for maintaining cache coherence (SGI, 2000a). Cache coherence is the ability of the computer system to maintain data consistency during parallel execution (to be discussed later in this section). Further, the processors in a node board are connected to the I/O devices through a hub.

The Origin 2000 multiprocessor used in this work at the Ohio Supercomputer Center has thirty-two 300 MHz IP27 processors arranged into 16 node boards (Figure 6.1). Each processor has an MIPS R12000 CPU, an MIPS R12000 floating-point unit, a 32 kbytes on-chip data cache, a 32 kbytes on-chip instruction cache, and an 8 Mbytes secondary unified cache (i.e. it serves as both data and instruction cache). Each two processors of a node board share a common local main memory with a size of 1.0 Gbytes. Thus, the 16 node boards have a total main memory of 16 Gbytes. Each processor can have access to the local main memories of other processors. The local main

Cost Optimization of Structures: Fuzzy Logic, Genetic Algorithms, and Parallel Computing   H. Adeli and K. C. Sarma   © 2006 John Wiley & Sons, Ltd

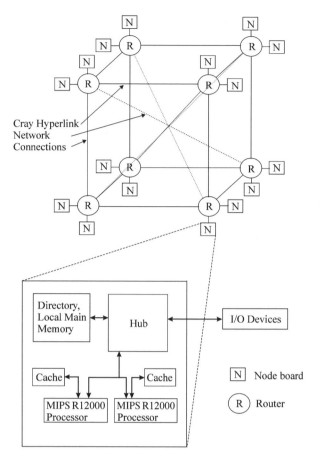

Figure 6.1 Architecture of the Origin 2000 supercomputer

memory of a processor will be the remote main memory of other processors on remote boards.

The distributed memories of the node boards are connected to other processors through routers via very fast Cray Hyperlink network connections, allowing efficient two-way information traffic among the connected processors. Though in the general hypercube configuration the processors are connected to their neighboring processors only (identified by solid lines in Figure 6.1), in the Origin 2000 supercomputer the most distant processors are also connected directly (shown by dotted lines in Figure 6.1) in order to increase the efficiency of the message flow.

For each processor the memory access hierarchy is (a) register (located inside the processor), (b) level-1 cache, (c) level-2 cache, (d) local main

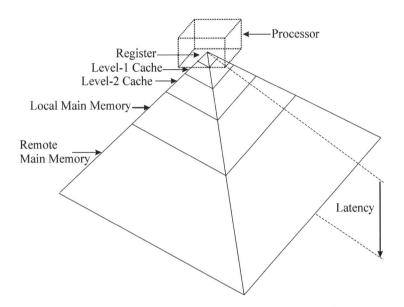

Figure 6.2 Pyramid representation of the memory hierarchy of Origin 2000

memory, and (e) remote main memory. Figure 6.2 shows the pyramidal nature of this memory hierarchy. It displays roughly the relative sizes of various memory types and the latency of retrieving data from them. The latencies of retrieving data from level-1 and level-2 caches are 1 and 10 clock periods (1 clock period = 3.33 nanoseconds on Origin 2000 used in this work), respectively. Latencies of memory access from local and remote main memories are 60 and [60 + 20*(number of router hops)] clock periods, respectively (Ennis and Baer, 1999). The data bandwidth on Origin 2000 is 600 Mbytes/s. Such a parallel processing computing environment is known as a scalable nonuniform memory access (NUMA) system. In such a system, instead of passing messages among the processors, the interconnections among the processors provide a distributed-shared memory that can be accessed by any processor but at different latencies.

By combining some hardware and software features, the operating system of Origin 2000 moves the most used segment (page) of read–write data to a memory closer to the processor. This feature is called page migration. In the case of read-only data, instead of migrating, the operating system replicates the data. By using these two features, page migration and replication, Origin 2000 reduces the latencies of memory access. For cache coherency,

Origin 2000 uses the directory-based coherency protocol. In this protocol, a processor takes exclusive ownership of the data when it modifies a cache line (a data segment on the cache) and invalidates the same data on other processors' caches by transmitting appropriate signals. Thus, other processors cannot use that particular cache line from their own cache to read the same data; instead, they obtain the data from the cache where the modified data exist. The cache, which owns that cache line, relinquishes its ownership once the data are updated in the main memory.

Though the Origin 2000 supercomputer at the Ohio Supercomputer Center has 32 processors, depending on the memory requirements and computational time only up to 16 processors are available to the users for parallel computation in a batch-processing environment.

6.2 Parallel Processing Implementation Environment

Since Origin 2000 is a distributed shared memory multiprocessor it supports both shared memory data parallel processing using OpenMP API (Application Programming Interface) and distributed memory message passing parallel processing using MPI (Message Passing Interface). In the next chapter, parallel algorithms are first created using both parallel processing environments and their relative efficiency is evaluated for large-scale structural optimization. Next, a bilevel parallel processing model is presented for large-scale structural optimization using GA through judicious combination of the two aforementioned approaches.

6.2.1 OpenMP Data Parallel Application Programming Interface (API)

OpenMP is an API that supports parallel programming in C/C++ and Fortran on many operating systems including UNIX and Windows NT, and is supported by many major hardware and software vendors (OpenMP, 2000). It consists of compiler directives, library routines, and environment variables (variables whose values are set in the shell environment independent of the application program) used for parallel computing in shared memory environments. The directives extend C/C++ and Fortran sequential programs to single-program multiple data (SPMD), work-sharing, and synchronization constructs. OpenMP facilitates portable and scalable parallel programming in a wide range of shared-memory multiprocessors. It is efficient both in fine-and course-grained parallelization. OpenMP can function efficiently in a globally addressable and cache coherent distributed-shared memory system with minimum latency and with no explicit address mapping such as Origin 2000. OpenMP provides many directives for managing private and shared data, query functions, and conditional compilation supports.

Using OpenMP directives, the parallel algorithm and software developer still has to manage data dependencies, racing conditions, and processor deadlocks that may cause incorrect parallel computations. OpenMP uses a fork-and-join paradigm of parallel execution (Figure 6.3). OpenMP creates a single thread, the so-called master thread, to begin the execution of a given parallel region. When it encounters parallel directives it creates additional threads, referred to as slave threads. At the end of the parallel region, the threads are joined and the computational results are passed on to the master thread. Each thread is created for a processor to perform a specific task. More than one thread can be created for any given processor.

For solving the sparse linear equations encountered in the analysis of structures, parallel banded solver routines (multiple-thread supported or thread-safe) from the package LAPACK (Anderson *et al.*, 1999) are used.

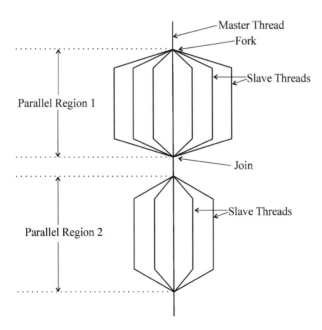

Figure 6.3 OpenMP fork-and-join parallel processing paradigm

6.2.2 Message Passing Interface (MPI)

Message Passing Interface (MPI) is the standard portable message passing parallel programming interface for distributed memory systems used by a large number of vendors, implementers, and end users since 1994. MPI is a library of message passing and related functions or routines in C and Fortran languages. It was originally created for numerically intensive computations on massively parallel computers and a networked cluster of workstations. It improves the performance of scalable multiprocessors with specialized interprocessor communication hardware such as Connection Machine and Cray T3E. It is now also available on shared-memory multiprocessors. The implementation of MPI on top of the standard UNIX interprocessor communication protocols enhances its portability in a heterogeneous network of computers. Though MPI has a large set of functions (in the order of 125), an effective message passing interface can often be created by using only its six basic functions. Language-independent notations are used to specify the functions of MPI. For parallel or distributed processing, MPI buffers messages temporarily and stores internal representations of different MPI objects without needing the parallel algorithm developer to specify their sizes or locations (MPI, 1995, 2000).

6.3 Performance Optimization of Parallel Programs

C programming language is used in the parallel processing work. The parallel processing performance of the implemented program is optimized using a number of schemes:

1. The number of function calls is reduced as much as possible without adversely affecting the modular structure of the parallel program.
2. Unit strides (index increments of arrays) are used as much as feasible in order to increase the rate of cache hits.
3. The memory allocation for arrays within functions (subroutines) is done statically in order to reduce the overhead for optimization of the executable code by the compiler. In the main program, however, this overhead is insignificant and consequently the memory is allocated dynamically.
4. Short loops are eliminated by repeating the expressions comprising the loops (loop unrolling) in order to eliminate the overhead associated with calling a loop.

5. Compiler directives are used to pre-fetch data to the cache before they are needed in the computation in order to reduce memory access time.
6. Small subroutines are transferred into larger subroutines (subroutine inlining) to eliminate procedure call overheads with the Origin-specific compiler directives.
7. On Origin 2000 compilers, a division operation takes more time than a multiplication operation; consequently, division operations are replaced by equivalent multiplication operations wherever possible.

7

Parallel Fuzzy Genetic Algorithms for Cost Optimization of Large Steel Structures

7.1 Genetic Algorithm and Parallel Processing

The advantage of the genetic algorithm (GA) in structural optimization is its improved chance of yielding the global optimum (Goldberg, 1989; Adeli and Hung, 1995). Adeli and Cheng (1993) presented an early application of GA in structural optimization, as discussed in Chapter 2. Many other structural engineering applications of GA have been published in recent years, e.g. Adeli and Cheng (1994a,1994b), Hajela *et al.* (1998), Jenkins (1998), Koumousis and Arsenis (1998), Lin and Yang (1998), Rajeev and Krishnamoorthy (1998), Soh and Yang (1998), Thierauf and Cai (1998), Park and Grierson (1999), Savic *et al.* (1999), and Yeh (1999).

The disadvantage of GA is the high cost of function evaluation, especially for large structures and larger population sizes, where many design iterations are needed to find the global optimum. In such cases, thousands of population strings have to be created, requiring a large number of structural analyses in each design iteration. In order to apply GA-based optimization to large structures Adeli and Cheng (1994b) were the first to present parallel or concurrent augmented Lagrangian genetic algorithms for the minimum weight design of space truss structures utilizing the multiprocessing capabilities

---

Cost Optimization of Structures: Fuzzy Logic, Genetic Algorithms, and Parallel Computing   H. Adeli and K. C. Sarma   © 2006 John Wiley & Sons, Ltd

of high-performance computers such as the Cray YMP8/864 (Adeli 1992a, 1992b; Adeli and Kamal, 1992, 1993; Adeli and Soegiarso, 1999; Adeli and Kumar, 1999). The parallel algorithms were used to find the minimum weight design of large space axial-force structures and their parallel processing speed-up was evaluated. For a 35-story space truss tower with 1262 members a high speed-up of 7.7 was achieved using eight processors.

An attractive feature of GA-based structural optimization algorithms is their adaptability to effective parallel processing on scalable distributed memory multiprocessors. In a GA-based optimization procedure 95 % to 98 % of the computation time is spent on the evaluation of fitness functions including the finite element analyses of structures (Adeli and Kumar, 1995b). The fitness function evaluation is done for each population separately. Therefore, in principle at least 95 % of the computations in the GA-based optimization algorithm can be performed concurrently without any need for interprocessor communications.

Adeli and Kumar (1995a) present a computational model and concurrent genetic algorithms for optimization (minimum weight design) of large space structures on massively parallel supercomputers by exploiting parallelism at both the coarse-grained design optimization level in the genetic search and the fine-grained fitness function evaluation level. The algorithms are implemented on Connection Machine CM-5 and applied to the minimum weight design of large steel structures, including a 4016-member tower, subjected to the constraints of the AISC ASD specifications (AISC, 1995). Adeli and Kumar (1995b) present a distributed GA for optimization of large structures on a cluster of workstations connected through a local area network.

As mentioned earlier, sequential execution of a GA-based optimization procedure for a large structure such as a moment-resisting space high-rise frame structure with thousands of members requires an excessive amount of processing time. This is an aspect of the GA hardly discussed in the literature because the published papers deal mostly with small to medium size problems. For example, the multi-criteria cost optimization of the 3228-member 36-story moment-resisting steel space frame with cross-bracings subjected to the AISC ASD code constraints mentioned earlier and reported in Chapter 5 takes a few days of a single processor of SGI Origin 2000. Further, solution of problems of this size using a GA-based optimization algorithm requires large cache sizes, a large main memory, and a high memory bandwidth (the rate of data communicated between the processors and the memory per second). Optimization of the aforementioned structure cannot be obtained on a single workstation in any reasonable amount of time.

This is the motivation for developing parallel algorithms on high-performance multiprocessor machines (Adeli 1992a,1992b; Adeli and Kamal, 1993; Adeli and Soegiarso, 1999).

7.2 Cost Optimization of Moment-Resisting Steel Space Structures

In Chapters 4 and 5 the basis of design was the AISC ASD code (AISC, 1995). In this chapter, both AISC ASD as well as the more complicated AISC LRFD code (AISC, 2001) are used. In a moment-resisting steel space frame, according to the LRFD code the following interaction equation has to be satisfied for each beam–column member i:

$$
\begin{aligned}
\frac{P_{ui}}{\phi_c P_{ni}} + \frac{8}{9}\left(\frac{M_{uix}}{\phi_b M_{nix}} + \frac{M_{uiy}}{\phi_b M_{niy}}\right) &\leq 1.0 \quad \text{for}\frac{P_{ui}}{\phi_c P_{ni}} \geq 0.2 \\
\frac{P_{ui}}{2\phi_c P_{ni}} + \left(\frac{M_{uix}}{\phi_b M_{nix}} + \frac{M_{uiy}}{\phi_b M_{niy}}\right) &\leq 1.0 \quad \text{for}\frac{P_{ui}}{\phi_c P_{ni}} < 0.2
\end{aligned}
\tag{7.1}
$$

where M_{uix} and M_{uiy} are the required flexural strengths of the ith member about the major and minor axes, respectively; M_{nix} and M_{niy} are the nominal flexural strengths of the ith member about the major and minor axes, respectively; P_{ui} and P_{ni} are the required and nominal compressive strengths of the ith member, respectively; and ϕ_b and ϕ_c are the resistance factors for flexure and compression, respectively. The nominal compressive strength of the ith member is calculated as

$$
P_{ni} = F_{cr} A_i
\tag{7.2}
$$

In equation (7.2), A_i is the cross-sectional area of the ith member and the critical buckling stress F_{cr} is calculated as

$$
F_{cr} = \begin{cases} (0.658^{\lambda_c^2})F_y & \text{for}\lambda_c \leq 1.5 \\ \left(\dfrac{0.877}{\lambda_c^2}\right)F_y & \text{for}\lambda_c > 1.5 \end{cases}
\tag{7.3}
$$

where

$$
\lambda_c = \frac{Kl_i}{\pi r}\sqrt{\frac{F_y}{E}}
$$

E is the modulus of elasticity, r is the governing radius of gyration, and F_y is the yield stress.

The AISC ASD and LRFD codes (AISC, 1995, 2001) provide alignment charts for finding the values of the effective length factor K based on the numerical solution of the transcendental equations. In this work, the values of K are found by approximate but explicit equations used in the European steel design code as defined by equations (5.6) and (5.7). It has to be pointed out that equation (7.1) is a complicated constraint and a highly nonlinear implicit and discontinuous function of design variables. Such constraints are known to create convergence difficulies for optimization algorithms.

7.3 Data Parallel Fuzzy Genetic Algorithm for Optimization of Steel Structures Using OpenMP

The GA-based optimization procedure is an intrinsically parallel method and may be implemented in a manner sometimes referred to as 'embarrassingly parallel', in the sense that all the processors may run in parallel without any need for communication. For example, for a string population of size N (say 200) and N_p processors (say 8), the workload can be divided equally by assigning each processor $N/N_p(200/8 = 25)$ populations. Therefore, perfect load balancing is achieved provided that N is a multiple of N_p. While this may be the best approach from a purely parallel processing point of view it will not create the most efficient GA-based optimization algorithm, especially for large systems. This is because the solution convergence of a GA-based optimization algorithm depends on the size of the string population.

A smaller population size usually requires more iterations for convergence to a local optimum. Further, by dividing a large population size into N_p smaller populations, the chance of obtaining the global optimum is decreased. Topping *et al.* (1998) report the ineffectiveness of the reproduction for smaller population sizes, 'resulting in incestuous crossbreeding and premature convergence'. However, a large population for a large structure consisting of thousands of members requires an inordinate amount of processing time. For a moment-resisting space structure (having six degrees of freedom per node) with thousands of members and several hundreds of design variables, a large population size in the order of two to three times the number of variables is needed.

The parallel fuzzy GA for cost optimization of steel structures based on OpenMP API is shown schematically in Figure 7.1. In this algorithm as well as the ones presented in the following two sections, parallel processing is performed only in the first stage of the cost optimization algorithm since the second stage takes less than 1 % of the total processing time. In the parallel

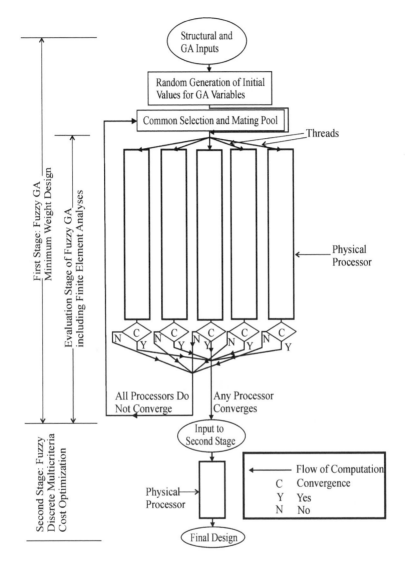

Figure 7.1 Macro flowchart for a data parallel fuzzy GA for optimization of steel structures using OpenMP

algorithm presented in Figure 7.1, only coarse grain parallel computations are performed. OpenMP parallel directives are used to create a number of threads to distribute the population among the processors equally. The population size is chosen as a multiple of the processors. The number of threads is equal to the number of assigned processors. Different processors of the Origin

2000 supercomputer have direct access to each other's memory, which is a prerequisite for data parallel processing with OpenMP directives.

Very large structures considered in this book require a large amount of memory in the order of 2 Gbytes or more for execution and storing the data. The capacity of each local memory on each node board of the Origin 2000 is only 1 Gbyte. In such a case, the data is distributed across the distributed memories of the Origin 2000. In a sequential processing, the single processor has to access data from the remote memories with non-uniform latencies, which slows down the processing speed substantially.

However, when OpenMP is used, instead of one processor, multiple processors will be acting on the dataset and each processor will have an easy access to its required data in its own local memory or in a memory in its neighborhood. Furthermore, the number of caches is also increased in proportion to the number of processors. There will be a fewer number of cache misses (when a processor does not find the data in a cache line and must reload the data). The page migration and replication features (discussed earlier) move or copy, respectively, the required data to the local memory of the processor or to the memories of processors close to that processor (when its own local memory is insufficient for the transferred data), resulting in a very high level of computational efficiency. Under OpenMP the Origin 2000 operating system tries to optimize the data locality for each processor. Due to these reasons, parallel processing using OpenMP results in a significant drop in computational time and super-linear speed-up. This is demonstrated in the speed-up results for the two large example structures presented in a subsequent section.

7.4 Distributed Parallel Fuzzy Genetic Algorithm for Optimization of Steel Structures Using MPI

For distributed message passing parallel processing using MPI two different schemes are investigated: the processor farming scheme and the migration scheme.

7.4.1 Processor Farming Scheme

The parallel fuzzy GA for optimization of steel structures based on the processor farming scheme is shown schematically in Figure 7.2. This algorithm is similar to the distributed GA of Adeli and Kumar (1995b) developed on a cluster of workstations using dynamic load balancing and message passing

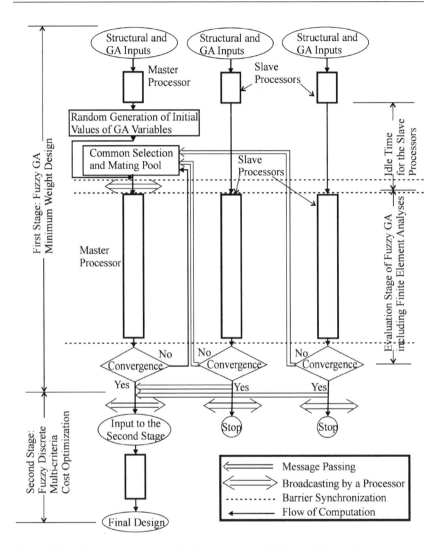

Figure 7.2 Macro flowchart for the distributed parallel fuzzy GA optimization of steel structures using the MPI and the processor farming scheme

constructs from the parallel virtual machine (PVM) API (Adeli and Kumar, 1999). Only coarse grain parallel computations are performed. MPI parallel function calls are used to define a master processor and $N_p - 1$ slave processors. The random generation of initial values of the GA variables for entire N populations is performed by the master processor. The master processor also performs the GA selection, crossover, and mutation operations for the second and subsequent generations.

The population size is chosen as a multiple of the total number of processors. For the evaluation stage of the fuzzy GA, which includes finite element analyses of the structure, N populations are divided among N_p available processors equally. At the end of this stage, the populations assigned to different processors are combined together in one common mating pool in the master processor, where GA selection, crossover, and mutation operations are performed. Consequently, the fitness function evaluation, which consumes at least 95 % of the total processing time, is done concurrently by all the processors.

The master processor initializes the population of strings and other genetic parameters. Then it broadcasts the information to the slave processors. The slave processors are active only at the fitness evaluation stage. In order to coordinate the communications among the slave and master processors, three barrier synchronization flags are inserted at different stages of computations. Each slave processor does its share of fitness function computations including the finite element analyses of the structure and sends its report (results) to the master processor through a message passing function. The master processor receives the reports from the slave processors one at a time (sequentially). Then, it initiates the GA selection, crossover, and mutation operations and broadcasts the new population information to the slave processors. The loop continues until one processor satisfies the convergence criteria. When a slave processor meets the convergence criteria it sends the final results to the master processor and broadcasts a signal to all other processors to stop. On the other hand, if the master processor meets the convergence criteria it only broadcasts the stopping signal to the slave processors. The broadcasting of the stopping signal to other processors is necessary to prevent the processors running unnecessarily and to avoid any deadlock situation.

7.4.2 Migration Scheme

In this scheme all processors are treated equally and N populations are divided among N_p available processors equally right from the beginning (Figure 7.3). Each processor randomly generates initial values of the GA variables and performs the GA selection, crossover, and mutation operations on its own assigned population. To improve the convergence of the algorithm a migration strategy is used to create diversity among the populations assigned to various processors. This diversity is achieved by creating a virtual ring topology for the N_p processors and passing a pre-set percentage (say, one-third or one-fourth) of populations from one processor to its neighbor

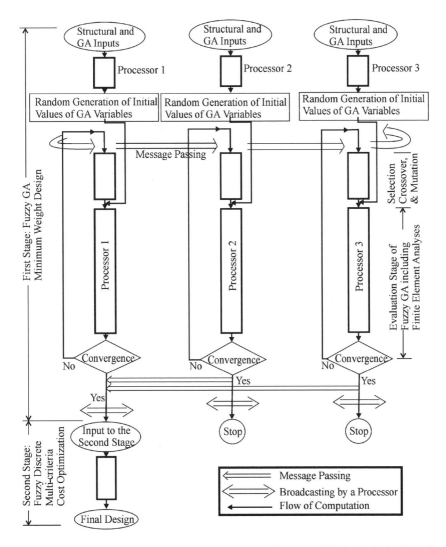

Figure 7.3 Macro flowchart for the distributed parallel fuzzy GA for optimization of steel structures using the MPI and the migration scheme

in a cyclic order. Each processor sends the pre-set percentage of population information to its right neighbor and receives the information for the same percentage of population from its left neighbor.

The MPI API provides the capability to create a virtual topology, where the operating system treats the participating processors according to a topology defined by the parallel algorithm and software developer. In this work, a one-dimensional virtual topology in the shape of a ring is created using the MPI virtual topology directive. In this scheme the processors are assigned some virtual coordinates (using numbers 0 to $N_p - 1$) to define neighboring processors in the virtual topology for the purpose of communication among themselves, even though physically they may not be neighbors. This scheme is scalable for any given number of processors, and consequently lends itself effectively to parallel processing.

For message passing each processor sends a message object with the information about the migrating population (according to the pre-set percentage) to its right neighbor and receives another object with similar information from its left neighbor (as defined by the virtual coordinates). These population objects contain the binary string of individual variables, their fitness values, and the number of population sent in the object. After a cycle of migration, each processor performs the fitness function evaluation on its complete diversified population. When a processor attains the Lagrangian function convergence criteria, it sends the results to processor 1 through a message passing construct and broadcasts a stopping signal to all other processors (Figure 7.3).

7.5 Bilevel Parallel Fuzzy GA for Optimization of Steel Structures Using OpenMP and MPI

In order to improve the parallel processing speed-up and efficiency of the algorithms, a bilevel parallel fuzzy GA is presented using OpenMP data parallel threads within the MPI distributed processing environment. The OpenMP threads/processors perform the fitness function evaluation using finite element analyses. The distributed shared memory architecture of the Origin 2000 supercomputer allows implementation of such a bilevel parallel algorithm, where parallel processing is performed at two levels with two different parallel processing paradigms, message passing and data parallel. Besides having direct access to each other's memory, which is a prerequisite for data parallel processing with OpenMP, Origin 2000 supports the message passing MPI libraries where each processor can have its own virtual private memory, which can be accessed by other processors only with message passing routines. The bilevel parallel algorithm, however, can be implemented on any distributed shared memory architecture.

Two different versions of the bilevel parallel algorithms are presented, one with processor farming and the other with migration message passing schemes. These two bilevel parallel algorithms are also presented schematically in Figures 7.4 and 7.5, respectively.

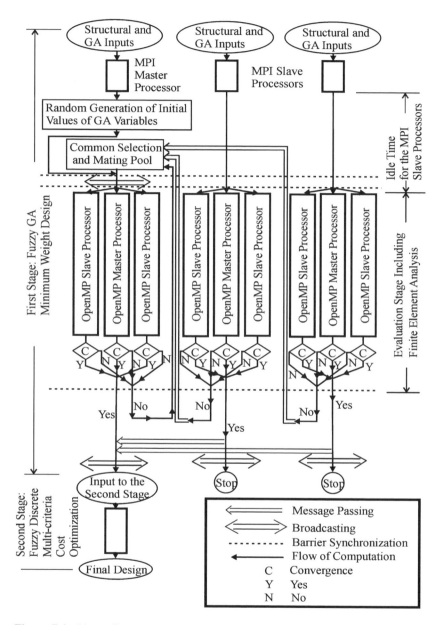

Figure 7.4 Macro flowchart for the parallel fuzzy GA for the optimization of steel structures using the bilevel method combining OpenMP and the MPI processor farming scheme

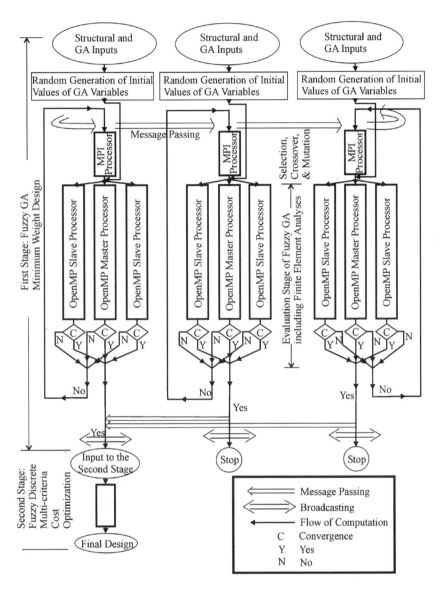

Figure 7.5 Macro flowchart for the parallel fuzzy GA for the optimization of steel structures using the bilevel method combining OpenMP and the MPI migration scheme

7.5.1 Bilevel Parallel Fuzzy GA with the Processor Farming Scheme

Step 1. Select a number of processors as MPI processors and initialize them. Processor 1 is called the master processor and the remaining processors are dubbed slave processors.

Step 2. Input the structural and GA attributes to the MPI processors concurrently.

Step 3. If the iteration number is one, generate the initial values of GA design variables randomly by the master processor, then go to step 5. Otherwise, go to step 4.

Step 4. Perform the genetic selection, crossover, and mutation operations using the MPI master processor.

Step 5. Broadcast the genetic information from the MPI master processor to the MPI slave processors (barrier synchronization flags are provided before and after the broadcasts to synchronize the slave processors with the master processor).

Step 6. Create multiple OpenMP threads or processors by MPI processors to perform evaluation of the fitness function using OpenMP directives. (The total number of threads is equal to the total number of active physical processors. The parallel processing control now goes to OpenMP.)

Step 7. Perform the fuzzy GA minimum weight design (the first stage of the multi-criteria cost optimization algorithm) including a fitness function evaluation, finite element analyses, an AISC design constraint evaluation, and computation of the fuzzy membership function by each OpenMP processor on its share of the population concurrently.

Step 8. Check the convergence criteria by each OpenMP processor. If the convergence criterion is not met, go to step 9. If the convergence criterion is met in an OpenMP processor, then the OpenMP loop is terminated in all OpenMP processors within the associated MPI processor and the control is transferred to that MPI processor. If the MPI processor is a slave processor it sends the results to the MPI master processor and broadcasts signals to other slave MPI processors to stop. Go to step 12.

Step 9. Join the OpenMP threads and release the OpenMP slave processors. (Barrier synchronization flags are provided after the OpenMP join operation so that each MPI processor waits for other MPI processors before sending the fitness function evaluation information to the master processor.)

Step 10. Send fitness function information from the slave processors to the master processor sequentially using the MPI send and receive procedures.
Step 11. If the iteration number is less than the specified maximum number of iterations go to step 4.
Step 12. Stop MPI.

7.5.2 Bilevel Parallel Fuzzy GA with the Migration Scheme

Step 1. Select a number of processors as MPI processors and initialize them.
Step 2. Input the structural and GA attributes to the MPI processors concurrently.
Step 3. If the iteration number is one, generate the initial values of GA design variables randomly by each MPI processor for its share of the N/N_p population and then go to step 5.
Step 4. Perform the genetic selection, crossover, and mutation operations by each MPI processor on its share of the N/N_p population.
Step 5. Create multiple OpenMP threads or processors by MPI processors to perform an evaluation of the fitness function using OpenMP directives. (The total number of threads is equal to the total number of active physical processors. The parallel processing control now goes to OpenMP.)
Step 6. Perform the fuzzy GA minimum weight design (the first stage of the multi-criteria cost optimization algorithm), including a fitness function evaluation, finite element analyses, an AISC design constraint evaluation, and computation of the fuzzy membership function by each OpenMP processor on its share of the population concurrently.
Step 7. Check the convergence criteria by each OpenMP processor. If the convergence criterion is not met, go to step 8. If the convergence criterion is met in an OpenMP processor, then the OpenMP loop is terminated in all OpenMP processors within the associated MPI processor and the control is transferred to that MPI processor. If the MPI processor is not processor 1, it sends the results to processor 1. The MPI processor broadcasts signals to other MPI processors to stop. Go to step 11.
Step 8. Join the OpenMP threads and release the OpenMP slave processors.
Step 9. Send and receive a predetermined percentage of the population information (for example, one-fourth) to and from the right and left neighboring processors, respectively, using the virtual ring topology of the MPI and MPI send-and-receive protocol.

Step 10. If the number of iterations is less than the specified number of
iterations go to step 4.

Step 11. Stop MPI.

7.6 Application to High-Rise Building Steel Structures

The implemented parallel algorithms are used to find the minimum cost
design solutions for two large moment-resisting steel space frame building
structures. These structures are designed according to both AISC ASD and
LRFD specifications (AISC, 1995, 2001).

7.6.1 Example 1

This example is the same 36-story irregular moment-resisting steel space frame
structure with setbacks and cross-bracings discussed in Chapter 5 (Figures 5.6
and 5.7). Details of this example and the loading and displacement constraints are presented in Adeli and Park (1998) and Chapter 5. Figure 7.6

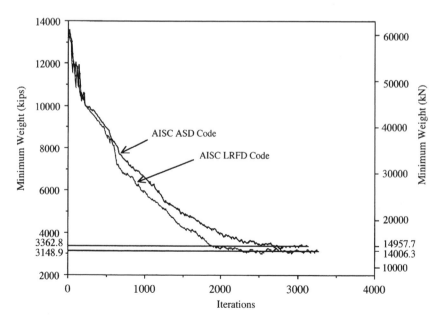

Figure 7.6 Design histories for the fuzzy GA minimum weight design for the 36-story
example structure using the MPI and the migration scheme

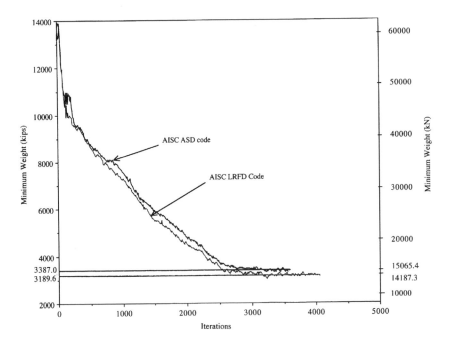

Figure 7.7   Design histories for the fuzzy GA minimum weight design for the 36-story example structure using the MPI and the processor farming scheme

presents the convergence results for the distributed fuzzy GA using the MPI and the migration scheme. Figure 7.7 presents the convergence results for the distributed fuzzy GA using the MPI with the processor farming scheme. The data parallel fuzzy GA using OpenMP would yield results similar to those presented in Figure 7.7. The optimum design solutions obtained by the fuzzy multi-criteria discrete optimization model based on the LRFD code for cases A, B, C, and D defined in Section 5.8 are presented in Table 7.1, together with the minimum weight solution reported in Adeli and Park (1998). Similar results based on the ASD code are presented in Table 5.2.

Using the LRFD code, the minimum material cost solution, case A of the current study, is 29.5 % less expensive than the minimum weight solution of Adeli and Park (1998), but the number of different types of shapes is 51 compared with 24 in Adeli and Park (1998). The increases of the material costs with the reduction in the number of section types by increasing the value of N_i are also studied. The optimum solution obtained for case C(iii) with only 23 different types of shapes is about 13.5 % less expensive than the minimum weight solution of Adeli and Park (1998) with 24 different types of shapes.

Table 7.1   Optimum design solutions obtained by the fuzzy discrete multi-criteria cost optimization model based on the LRFD code for the 36-story structure (Example 1)

Type of study	Material cost of the structure	Weight of the structure	Total number of section types
Case A for $N_i = 6, 10, 15$	$6 133 422	14538.3 kN (3268.5 kips)	51
Case B for $N_i = 6, 10, 15$	$6 265 238	14470.2 kN (3253.2 kips)	60
Case C(i) for $N_i = 6$	$6 321 856	14959.1 kN (3363.1 kips)	36
Case C(ii) for $N_i = 10$	$6 789 793	15924.3 kN (3580.1 kips)	28
Case C(iii) for $N_i = 15$	$7 525 915	17517.6 kN (3938.3 kips)	23
Case D(i) for $N_i = 6$	$6 279 168	14902.1 kN (3350.3 kips)	37
Case D(ii) for $N_i = 10$	$6 576 508	15574.7 kN (3501.5 kips)	29
Case D(iii) for $N_i = 15$	$7 012 014	16403.8 kN (3687.9 kips)	24
Adeli and Park (1998) (discrete minimum weight design)	$8 696 327	20342.9 kN (4573.5 kips)	24

Comparing the optimum design results based on the LRFD code presented in Table 7.1 with similar results based on the ASD code presented in Table 5.2, it is observed that the LRFD code results in cost savings in the range of 3.0 % (for the minimum section type solution) to 6.9 % (for the minimum weight solution).

7.6.2 Example 2

This example (Figures 7.8 and 7.9) is a 144-story super high-rise multi-story steel building frame structure classified as a modified tube-in-tube system with moment-resisting space frames and exterior cross-bracings. A minimum weight solution for the same example is presented by Adeli and Park (1998) who created and solved this large-scale optimization problem for the first time (Park and Adeli, 1997a). This structure has 8463 joints and 20 096

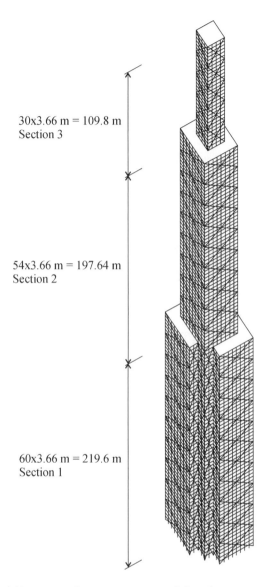

30x3.66 m = 109.8 m
Section 3

54x3.66 m = 197.64 m
Section 2

60x3.66 m = 219.6 m
Section 1

Figure 7.8 A 144-story steel space moment-resisting frame structure (Park and Adeli, 1997a)

members linked to 568 design variables, using the same design linking strategy used by Adeli and Park (1998). The structure is designed for combinations of dead, live, and wind loads according to the Uniform Building Code (1997). For details of this example including the loading the reader

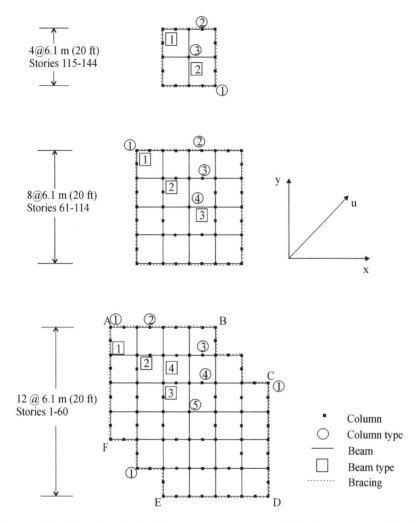

Figure 7.9 Floor plan for the 144-story space moment-resisting frame structure (Park and Adeli, 1997a)

should refer to Adeli and Park (1998). Figure 7.10 presents the convergence results for the distributed fuzzy GA using the MPI and the migration scheme. Figure 7.11 presents the convergence results for the distributed fuzzy GA using the MPI with the processor farming scheme. The data parallel fuzzy GA using OpenMP would yield results similar to those presented in Figure 7.11.

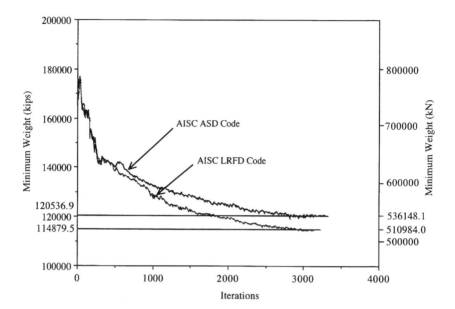

Figure 7.10   Design histories for the fuzzy GA minimum weight design for the 144-story example structure using the MPI and the migration scheme

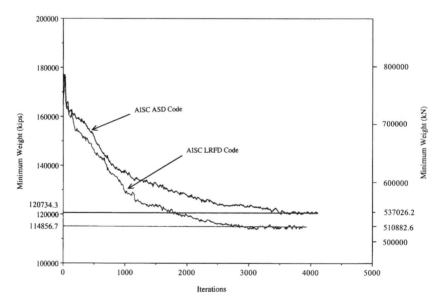

Figure 7.11   Design histories for the fuzzy GA minimum weight design for the 144-story example structure using the MPI and the processor farming scheme

The optimum design solutions obtained by the fuzzy multi-criteria discrete optimization model based on the ASD and LRFD codes for cases A, B, C, and D defined in Section 5.8 are presented in Tables 7.2 and 7.3, respectively, along with the minimum weight solutions reported in Adeli and Park (1998).

Table 7.2 Optimum design solutions obtained by the fuzzy discrete multi-criteria cost optimization model based on the ASD code for the 144-story structure (Example 2)

Type of study	Material cost of the structure	Weight of the structure	Total number of section types
Case A for $N_i = 6, 10, 15$	$281 265 784	544 950.7 kN (122 515.9 kips)	67
Case B for $N_i = 6, 10, 15$	$284 129 192	543 004.3 kN (122 078.3 kips)	97
Case C(i) for $N_i = 6$	$295 334 062	573 030.5 kN (128 828.8 kips)	49
Case C(ii) for $N_i = 10$	$310 848 325	604 287.5 kN (135 856.0 kips)	34
Case C(iii) for $N_i = 15$	$333 705 045	648 279.5 kN (145 746.3 kips)	27
Case D(i) for $N_i = 6$	$295 296 654	572 684.0 kN (128 750.9 kips)	52
Case D(ii) for $N_i = 10$	$302 604 951	587 969.1 kN (132 187.3 kips)	39
Case D(iii) for $N_i = 15$	$331 035 636	642 985.5 kN (144 556.1 kips)	31
Adeli and Park (1998) (discrete minimum weight design)	$367 139 910	682 240.5 kN (153 381.4 kips)	20

Table 7.3 Optimum design solutions obtained by the fuzzy discrete multi-criteria cost optimization model based on the LRFD code for the 144-story structure (Example 2)

Type of study	Material cost of the structure	Weight of the structure	Total number of section types
Case A for $N_i = 6, 10, 15$	$269 325 529	520 721.1 kN (117 068.6 kips)	69
Case B for $N_i = 6, 10, 15$	$272 036 160	519 180.8 kN (116 722.3 kips)	93
Case C(i) for $N_i = 6$	$291 505 329	564 522.4 kN (126 916.0 kips)	54

Table 7.3   (Continued)

Type of study	Material cost of the structure	Weight of the structure	Total number of section types
Case C(ii) for $N_i = 10$	$311 123 932	604 678.0 kN (135 943.8 kips)	35
Case C(iii) for $N_i = 15$	$328 089 790	636 515.9 kN (143 101.6 kips)	33
Case D(i) for $N_i = 6$	$280 182 608	541 977.7 kN (121 847.5 kips)	59
Case D(ii) for $N_i = 10$	$295 543 749	573 461.1 kN (128 925.6 kips)	49
Case D(iii) for $N_i = 15$	$309 322 213	599 830.6 kN (134 854.0 kips)	37
Adeli and Park (1998) (discrete minimum weight design)	$359 879 363	669 278.1 kN (150 467.2 kips)	21

Case B of the current study yields the minimum weight solution. As in Adeli and Park (1998), in this example standard wide flange rolled shapes cannot be used for the columns in section 1, some of the columns in section 2, and a few beams at the top of section 1. Built-up sections consisting of a W shape and two plates are used for those members, as shown in Figure 7.12. The costs of these built-up sections are assumed to be proportional to the weight of the original rolled shapes on which they are fabricated.

For an optimum design based on the ASD code, the minimum material cost solution, case A of the current study, is 23.4 % less expensive than the minimum weight solution of Adeli and Park (1998), but the number of different types of shapes is 67 compared with 20 in Adeli and Park (1998). Cases C and D of the current study present the solution for the minimum section type and equal importance of the three criteria. The increases of the material costs with the reduction in number of section types by increasing the value of N_i are also studied for these two cases and are noted in Table 7.2. The optimum solution obtained for case C(iii) is about 9.1 % less expensive than the minimum weight solution of Adeli and Park (1998). Similar results are obtained for optimum designs based on theLRFD code.

Comparing the optimum design results based on the ASD code presented in Table 7.2 with those based on the LRFD code presented in Table 7.3, it is observed that the latter code results in cost savings in the range of 1.7 %

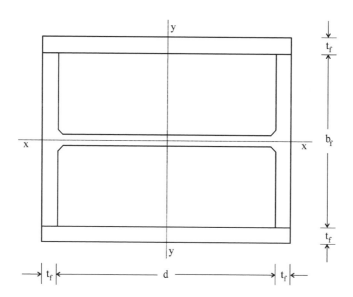

Figure 7.12 Built-up section used in the 144-story steel space moment-resisting frame structure

(for the minimum section type solution) to 4.3 % (for the minimum weight solution).

7.7 Parallel Processing Performance Evaluation

7.7.1 Data Parallel Fuzzy GA Using OpenMP

Speed-up results for Example 1 (the 36-story structure) relative to two processors for up to 16 processors (eight-node boards) are presented in Figure 7.13. It was not possible to obtain converged results using a single processor within a reasonable time. Thus, the speed-up is measured compared with the wall-clock time using two processors. The 36-story structure requires about 2 Gbytes of main memory for execution of the code. As explained earlier in this chapter, and shown in Figure 7.13, the data parallel algorithm using OpenMP results in a super-linear speed-up curve. It is observed in Figure 7.13 that the speed-up curve reaches its peak value of 3.06 for six processors, after which it declines. Speed-up reduces to 2.88 for eight processors and remains virtually the same (2.80) for up to 16 processors.

This can be explained as follows. Six processors (three node boards) providing 3 Gbytes of memory accommodate approximately 2 Gbytes of

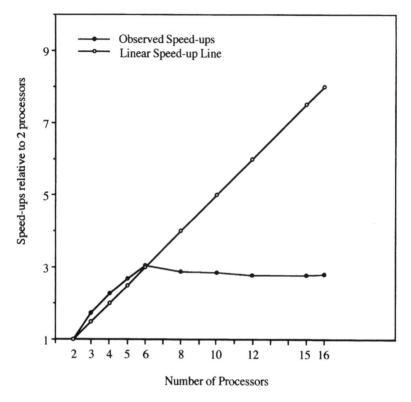

Figure 7.13   Speed-up curve for the OpenMP data parallel technique for the 36-story example structure

memory needed for this example, and allow the data parallel OpenMP API to access data efficiently, resulting in high-level speed-ups. Beyond six processors there is no further gain in memory access efficiency and parallel processing speed-up.

For problems requiring more than 8 Gbytes of memory, the number of Origin 2000 processors available at the Ohio Supercomputer Center is limited to a maximum of eight. Speed-up results for Example 2 (the 144-story structure) relative to three processors using up to eight processors (four-node boards) are presented in Figure 7.14. The reason for measuring speed-up relative to three processors is that at least three processors are needed for obtaining converged results for the 144-story structure. This structure requires about 10 Gbytes of main memory for execution of the code. As in Example 1, the data parallel algorithm using OpenMP results in a super-linear speed-up curve. In Figure 7.14, the speed-up increases with the increase in the number

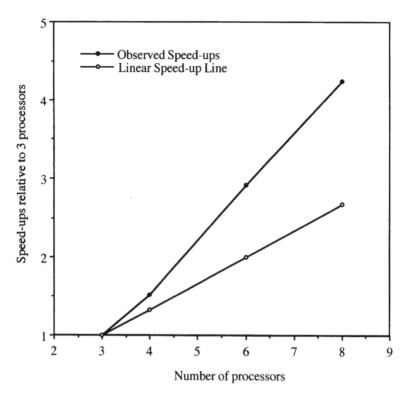

Figure 7.14 Speed-up curves for the OpenMP data parallel technique for the 144-story example structure

of processors, reaching a peak value of 4.24 for eight processors. This high value of speed-up using OpenMP is due to the tremendous amount of cache misses and page faults (when the data are not found in the main memory and have to be copied from the disk) encountered in sequential operation of a very large dataset. The number of cache misses and page faults reduce drastically as the number of processors is increased and the Origin 2000 operating system moves and copies the data to the memories closer to the processors by using page migration and replication procedures.

7.7.2 Distributed Parallel Fuzzy GA Using MPI

In distributed parallel processing using MPI, a processor cannot access the memory of any other processor except with a request to that effect using an MPI function call, which is a communication bottleneck creating a rather significant overhead. Compared with OpenMP, the parallel processing performance of a parallel algorithm using the MPI is not impressive due to the nonproximity of available data, unlike OpenMP processing where the operating system tries to optimize the data locality.

7.7.2.1 Processor Farming Scheme

In this scheme, the slave processors are at work in the fitness function evaluation stage and idle during the remaining operations (2 % to 5 % of the total computation time), resulting in deviation from the ideal parallel processing speed-up. Speed-up results for Example 1 (the 36-story structure) relative to two processors using up to 16 processors are presented in Figure 7.15. Speed-up results for Example 2 (the 144-story structure) relative to three processors using up to eight processors are presented in Figure 7.16. The reasons for using two and three processors in the two examples are explained

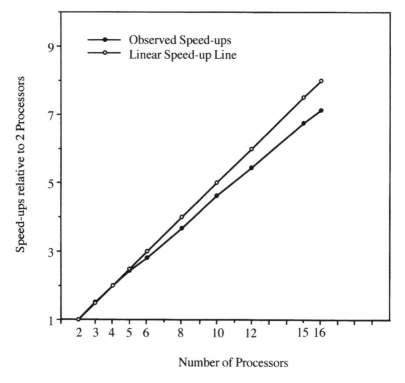

Figure 7.15 Speed-up curves for the MPI in the processor farming scheme for the 36-story example structure

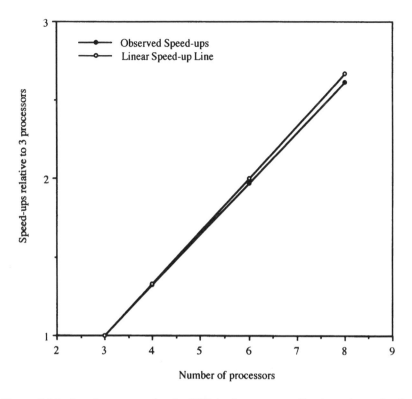

Figure 7.16   Speed-up curves for the MPI in the processor farming scheme for the 144-story example strcuture

in the previous section. The speed-up curve in both figures are sublinear. The sublinearity of the parallel processing speed-up can be explained by the fact that the communication bottleneck and the associated overhead increase as the number of processors is increased. Message broadcasting, introduction of a few barrier synchronization flags, and sequential transfer of data from the slave processors to the master processor all contribute to the reduction of the parallel processing speed-up as a function of the number of processors.

7.7.2.2 Migration Scheme

Speed-up results for Example 1 (the 36-story structure) relative to two processors using up to 16 processors are presented in Figure 7.17. Speed-up results for Example 2 (the 144-story structure) relative to three processors using up to eight processors are presented in Figure 7.18. It is observed that speed-up is slightly super-linear as in this scheme each processor performs

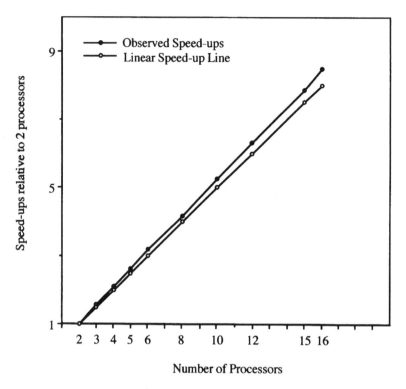

Figure 7.17   Speed-up curves for the MPI in the migration scheme for the 36-story example structure

genetic selection, crossover, and mutation operations on the same but smaller population sizes. Furthermore, an asynchronous message passing function is used for transmitting one-fourth of the populations to the neighbors with no need for any synchronization for migrating the populations. The overhead for sending a message asynchronously is much less than sending it with synchronization. The reason for migration is to have some diversity in the populations, but the time of migration is not crucial. Thus, the processors do not have to wait for passing or receiving messages. The processor involved accesses a smaller dataset with more data locality (closer to the processor).

7.7.3 Bilevel Parallel Fuzzy GA Using OpenMP and MPI

The bilevel parallel algorithm utilizes the best features of both MPI and OpenMP. As seen from Figure 7.13, the OpenMP procedure is efficient only

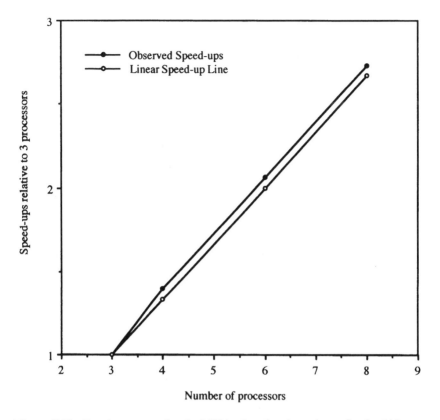

Figure 7.18   Speed-up curves for the MPI in the migration scheme for the 144-story example structure

when the local memory contains all the required data. The GA populations are distributed among different MPI processors. When OpenMP parallel constructs are used within an MPI processor, a smaller population is assigned to each OpenMP processor. The number of OpenMP processors chosen is in the range of 2 to 8. The data corresponding to the smaller population are closer to this new group of OpenMP processors and their data locality is much better than any other arrangements. The processors operate on the local data in a shared memory environment with minimum latency. Thus, the speed-up results of this bilevel scheme are much better than either OpenMP or MPI parallel processing procedures. Figures 7.19 and 7.20 show the comparative bar charts of speed-ups using different combinations of processors for the OpenMP and MPI procedures, using processor farming and the migration schemes, respectively, for Example 1. The speed-ups are

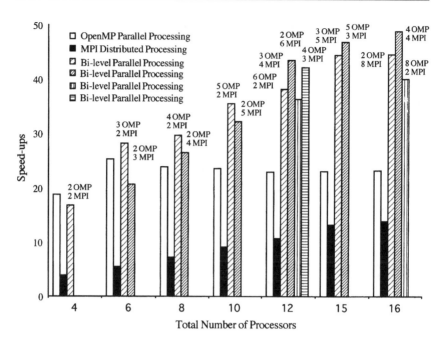

Figure 7.19 Comparative study of the bilevel algorithm with different combination of processors using the processor farming scheme for the 36-story example structure

calculated with respect to one processor by extrapolating the processing time of a single processor to complete the convergence. This is done to have a common speed-up measuring scale for all the three different parallel processing techniques. In these figures, OMP and MPI on the top of the bars refer to numbers of OpenMP and MPI processors, respectively. Note that the product of these two numbers is always equal to the total number of allocated processors.

In general, the speed-ups for the bilevel algorithms are higher than those of the OpenMP. The results for the bilevel algorithm with the migration scheme are better than those with the processor farming scheme. Using 16 processors, the speed-ups of the bilevel algorithm with the migration scheme is 55.9 compared with 23.1 for the OpenMP alone. Furthermore, the total convergence time of the fuzzy GA optimization procedure with the bilevel migration scheme is 23.3 % less than that with the bilevel processor farming scheme with eight processors (two MPI and four OpenMP processors). With the same number of processors, the total convergence time with the bilevel migration scheme is 38.6 % less than that with the OpenMP scheme.

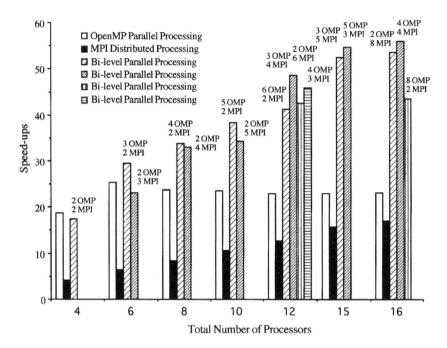

Figure 7.20 Comparative study of the bilevel algorithm with different combination of processors using the migration scheme for the 36-story example structure

For four processors the speed-up of the OpenMP parallel algorithm is slightly higher than the bilevel parallel algorithm, because data are available in the local memory of the small number of processors with minimum latency and without any need for message passing and its required overhead. However, with the increase in the total number of processors the speed-up of the bilevel parallel algorithm increases. From Figures 7.19 and 7.20 it is observed that the best results are obtained for 16 processors with four MPI and four OpenMP processors (55.9). The speed-ups of the MPI processors usually suffer from MPI message passing overheads and therefore their number should not be more than the number of OpenMP processors in each group. Similarly, the OpenMP processors are efficient only when the data they are working on are close to the processors and their best number is four. This is because in the hypercube structure of Origin 2000 there are four processors in each of the eight nodes (Figure 6.1). In this work Example 2 could not be tested with the bilevel parallel algorithm effectively as a sufficient number of processors (at least 16) was not available at the Ohio Supercomputer Center to work on a dataset, which requires about 10 Gbytes of memory.

7.8 Concluding Comments

The bilevel parallel algorithms for the fuzzy genetic optimization of large structures is effective in reducing the computational time on a cache-coherent nonuniform memory access (ccNUMA) parallel computer. The computational efficiency depends on the number of parallel processors, the combination of MPI and OpenMP controlled parallel processors, and the memory requirements. The present algorithm is effective in reducing the computational time significantly in comparison to the data parallel procedure using OpenMP API for the following two conditions: (1) if the memory is distributed among the main memories on the node boards across the processors and (2) if the number of available processors is sufficiently large in comparison to the node boards, preferably more than double. Furthermore, the migration scheme is more effective than the processor farming scheme for reducing the computational time for convergence of the optimum design and enhancing the performance of parallel processing.

8

Life-Cycle Cost Optimization of Steel Structures

8.1 Introduction

Life-cycle cost is the total cost of a structure during its lifetime. The life-cycle cost of a structure includes the initial costs including the costs of design and construction plus the costs of operation (utilities), maintenance (including repair), and eventually dismantling or demolishing the structure over the lifetime of the structure. If the initial cost of a structure is low but the utilities and maintenance costs are high, the structure may not be considered the most economical design. From an economic point of view the ideal goal of the cost optimization of structures should be minimizing the total life-cycle cost.

The three major costs during the life of a structure are the initial design/build cost, the operating cost, and the maintenance cost. The initial cost includes the costs of mechanical services, electrical services, finishing, interior and exterior decorations, external and internal facilities, and landscaping. The share of design and construction costs of the super- and substructures can be less than 50% of the total initial cost of the structure (Tietz, 1987).

The operating cost of the structure includes the costs of caretaking and cleaning, energy (gas and electricity), water and sewerage, insurance, mortgage interests, and security and management. The maintenance cost includes inspection, painting, repair, and replacement costs. The maintenance cost

Cost Optimization of Structures: Fuzzy Logic, Genetic Algorithms, and Parallel Computing H. Adeli and K. C. Sarma © 2006 John Wiley & Sons, Ltd

varies with the use, utility, and importance of the structure. Considering a discount rate (time value of money) of 2 % above inflation, a 50-year-old building usually has an operating cost 0.8 to 1.3 times the initial cost (Tietz, 1987).

The design life (or economic life) of most structures is often in the range of 30 to 40 years. However, the anticipated life (or actual life expectancy) of a structure is much longer; in the US it may be 60 to 85 years or even more (Tietz, 1987). The anticipated life of a structure is another important factor in the life-cycle design, but some other factors may also influence the anticipated life of a structure, like obsolescence, a natural or man-made catastrophe, and inadequate and out-of-fashion facilities.

The life expectancy as well as repair and maintenance also depend on the type of material used in the structure. Concrete loses its strength with the passing of time. Frequent changes in weather, cracks, shrinkage, and corrosion of reinforcements may reduce the life of a concrete structure. In these structures, even if repair is done, the initial strength may not be achieved. Similarly, in steel structures, the joints are the most vulnerable

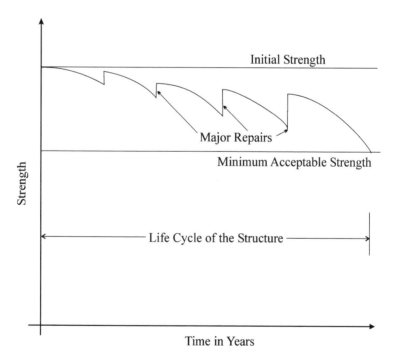

Figure 8.1 A pictorial view of the effect of repair on the strength of a structure with the passing of time

points. In an exposed structure the joints may accumulate dirt and debris, and if proper maintenance is not done this may lead to corrosion or rusting, leading to failure. In this respect riveted or bolted joins are inferior to welded joints. Figure 8.1 presents a pictorial view of the effect of repair on the strength of a structure with the passing of time.

8.2 Life-Cycle Cost of a Steel Structure and the Primary Contributing Factors

The life-cycle cost of a steel structure can be considered as the sum of seven different cost components or functions:

(1) initial cost, which includes the nine different costs including the material cost as defined in Section 5.2;
(2) maintenance cost, such as painting of exposed members of a steel structure;
(3) inspection cost to prevent potentially major damage to the structure;
(4) repair cost;
(5) operating cost required for proper functional use of the structure, such as heating and electricity;
(6) probable failure cost, based on the probability of failure;
(7) dismantling or demolishing costs.

Figure 8.2 presents different cost functions and their relationships in contributing to the life-cycle cost. Eleven main factors are identified that influence the life-cycle cost of a structure significantly. These are:

(a) cost of the rolled sections used for initial construction of the structure;
(b) the number of different section types used in the structure;
(c) the weight of rolled sections used in the structure;
(d) the surface area of rolled sections in the structure;
(e) the number of connections;
(f) the geographic location of the project site;
(g) the maintenance policy of the structure;
(h) the anticipated life of the structure;
(i) the discount rate of the currency;
(j) the use of the structure; and
(k) the importance of the structure.

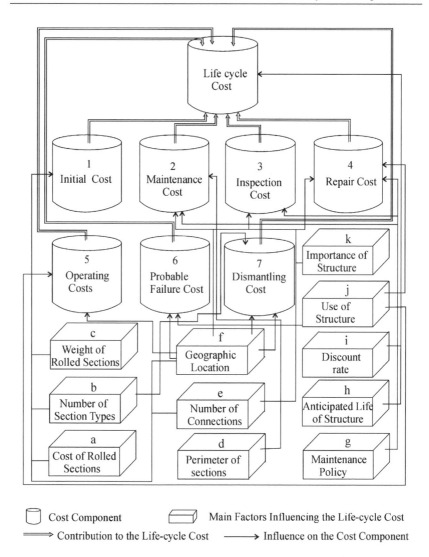

Cost Component   Main Factors Influencing the Life-cycle Cost

⟹ Contribution to the Life-cycle Cost   ⟶ Influence on the Cost Component

Figure 8.2 Different cost functions and their relationship in contributing to the life-cycle cost

The influence of these factors on different cost functions is shown in Figure 8.2 with single lines.

Out of the aforementioned 11 factors, factors (a), (b), (c), (e), and (f) are responsible for reducing cost function 1, as noted in Section 5.2. Factor (b) also influences cost function 7. Factor (d) is responsible for reducing cost function 2 as the cost of painting on an exposed steel structure for preventing possible rusting and corrosion depends primarily on the surface area of the members where paint is applied. Factor (e) influences cost function 1 in the form of fabrication, erection, connection materials, and labor costs. This factor also influences cost functions 2, 3, and 4, because the connections are usually the most vulnerable points for failure and have to be inspected periodically. Furthermore, painting of connections is usually time consuming and consequently costs more than the rest of the structure.

Factor (f) influences cost functions 1, 2, 3, 4, 5, and 6. Apart from influencing the initial cost, this factor influences the painting cost because the site may be near to the seacoast or in an industrially polluted area where corrosion is high. Geographic location also influences the maintenance and repair costs as a place with an abundance of skilled and unskilled labor force costs less than a place where the labor force is scarce and expensive. Maintenance and repair costs of structures in a difficult terrain are often expensive. Geographic location also influences the operating cost of a structure such as heating and air-conditioning. The probable failure cost also depends on the geographic location. For example, the probability of failure of a structure in an active earthquake zone increases significantly. Factor (g) influences cost functions 2, 3, 4, and 6. A poor maintenance policy often leads to early failure but a conservative maintenance policy, on the contrary, may result in excessive costs. Factors (h) and (i) influence all the cost functions, except cost function 1, and will be explained in the next section. Factor (k) influences cost functions 2, 3, 4, and 6. For very important structures like a nuclear power plant the maintenance and probable failure costs are high; consequently such structures are designed with a low probability of failure.

In this chapter, only the first four factors will be considered, as the structural designer has no control over the other seven factors to reduce the life-cycle cost of a structure. Extending the work presented in Chapter 5, a four-criteria cost optimization model is presented for the life-cycle cost optimization of steel structures. These criteria are (a) select discrete commercially available sections with the lowest cost, (b) select discrete commercially available sections with the lightest weight, (c) select the minimum number of different types of discrete commercially available sections, and (d) select discrete commercially available sections with a minimum total perimeter length for a minimum surface area, which determines the maintenance cost of exposed steel structures in sports utilities, bridges, towers, etc.

8.3 Formulation of the Total Life-Cycle Cost

The life-cycle cost of a steel structure can be formulated as follows:

$$
C_{\text{Lifecycle}} = C_{\text{Initial}} + \sum \frac{1}{(1+i)^{y_{n_1}}} C_{\text{Maintenance}} + \sum \frac{1}{(1+i)^{y_{n_2}}} C_{\text{Inspection}}
$$

$$
+ \sum \frac{1}{(1+i)^{y_{n_3}}} C_{\text{Repair}} + \sum \frac{1}{(1+i)^{y_{n_4}}} C_{\text{Operating}}
$$

$$
+ \frac{1}{(1+i)^{y_{n_5}}} C_{\text{Failure}} + \frac{1}{(1+i)^{y_{n_6}}} C_{\text{Dismantle}} \tag{8.1}
$$

where $C_{\text{Lifecycle}}$, C_{Initial}, $C_{\text{Maintenance}}$, $C_{\text{Inspection}}$, C_{Repair}, $C_{\text{Operating}}$, C_{Failure}, and $C_{\text{Dismantle}}$ are the total life-cycle, initial, maintenance, inspection, repair, operating, probable failure, and dismantling costs of a steel structure, respectively; i is the discount rate of money; y_{n_1}, y_{n_2}, y_{n_3}, y_{n_4}, y_{n_5}, and y_{n_6} are the years when the associated costs incur; and $\sum$ denotes the summation of all the costs of the same category during the life of the structure. Equation (8.1) is based on the concept of 'single present worth' (Kirk and Dell'Isola, 1995). The cost terms in the right-hand side of equation (8.1) are the costs in the year they actually occur. The $1/(1+i)^n$ factor is used to convert the cost into its present value discounted by the discount rate of i for the y_n period.

The discount rate depends on the prevailing interest rate and the depreciation of the currency or inflation rate. This rate is not a constant term and may vary over the life of the structure. A discount rate of 2% or 3% above inflation is considered an appropriate value (Tietz, 1987). Unfortunately, actual cost data needed in the life-cycle optimization of a structure are virtually nonexistent in the literature. The scant information available is based on insufficient statistical data or assumptions. Therefore, it is currently not feasible to optimize the life-cycle cost of a structure using equation (8.1) due to the lack of statistically meaningful cost data.

Wilson *et al.* (1997) discuss the development of a decision support system for analyzing the life-cycle costs of alternative bridge deck designs. A few researchers have presented probabilistic models for life-cycle cost optimization of structures (Frangopol *et al.*, 1997; Koskisto and Ellingwood, 1997). The use of such probabilistic models is also limited due to the lack of statistical data.

In this chapter a life-cycle cost optimization model is presented based on fuzzy logic (Zadeh, 1965) with the goal of formalizing the life-cycle design process but with some input from the design engineer. The model is described by considering only one representative type of cost incurred over the life of the structure, i.e. preventive maintenance in the form of periodic painting of an exposed steel structure to avoid corrosion or rusting. Thus, in this chapter only two kinds of costs are considered: initial cost and the maintenance cost in the form of painting.

8.4 Fuzzy Discrete Multi-criteria Life-Cycle Cost Optimization

Following the model presented in Chapter 5, four fuzzy functions are defined: the material cost of the structure $\tilde{C}(\tilde{y})$, the weight of the structure $\tilde{W}(\tilde{y})$, the number of different section types $\tilde{T}(\tilde{y})$, and the total surface areas of all the sections $\tilde{P}(\tilde{y})$ in terms of the fuzzy discrete variables (commercially available W shapes) $\tilde{y}$. The surface area of each member of a structure can be obtained as a product of the perimeter of the member section and the length of the member. These surface areas are important for calculating the painting cost of the structure.

The above-mentioned functions are fuzzy functions (identified by the sign $\sim$ on the top) because the variables $\tilde{y}$ are treated as fuzzy variables. The objective of the four-criteria optimization is to minimize the functions $\tilde{C}(\tilde{y})$, $\tilde{W}(\tilde{y})$, $\tilde{T}(\tilde{y})$, and $\tilde{P}(\tilde{y})$. Of these four fuzzy functions, $\tilde{C}(\tilde{y})$, $\tilde{W}(\tilde{y})$, and $\tilde{P}(\tilde{y})$ can be expressed explicitly in terms of the fuzzy variables $\tilde{y}$. The expressions for $\tilde{C}(\tilde{y})$ and $\tilde{W}(\tilde{y})$ are given by equations (5.1) and (5.2) respectively. The expression for $\tilde{P}(\tilde{y})$ is presented below:

$$\tilde{P}(\tilde{y}) = \sum_{i=1}^{N_t} l_i \tilde{y}_{p_i}, \qquad \tilde{y}_{p_i} \in S_{p_i} \tag{8.2}$$

where $\tilde{y}_{p_i}$ is the cross-sectional perimeter of the discrete standard shape with the maximum membership function corresponding to the minimum surface area criterion, l_i is the total length of members linked to variable x_i (representing the cross-sectional area of the ith group of members linked together in the continuous variable fuzzy GA optimization stage), S_{P_i} is a set of fuzzy discrete candidate standard shapes for the design variable x_i corresponding to the minimum perimeter criterion, and N_t is the number of initial section types (equal to the number of design variables in the first continuous variable stage of the optimization). The following approximate equation is used for the perimeter of wide flange shapes:

$$y_{p_i} = 4b_{fi} - 2t_{wi} + 2d_i \qquad (8.3)$$

where b_{fi}, t_{wi}, and d_i, are the flange width, web thickness, and the total depth of the section i.

In the fuzzy discrete multi-criteria cost optimization model, for each continuous variable $x_i (i = 1, 2, \ldots, N_t)$, N_i commercially available sections ($\tilde{y}_{ij}$) are selected such that $\tilde{y}_{ij} \geq x_i (j = 1, 2, \ldots, N_i)$. The commercially available sections, $\tilde{y}_{ij}$ have four fuzzy attributes corresponding to the four fuzzy objectives: minimizing material cost, minimizing weight, minimizing the number of section types, and minimizing the total surface areas of sections. All the N_i shapes selected are graded with membership functions for each of the four fuzzy criteria. The expressions for the membership functions for minimum cost, minimum weight, and minimum section type are presented in Chapter 5 (equations (5.12), (5.13), and (5.14)). Denoting the cross-sectional perimeter of the jth candidate section by p_j and the maximum and the minimum values within the N_i candidate shapes by $p_{j_{max}}$ and $p_{j_{min}}$, a linear membership function is defined in the following form (Figure 8.3):

$$\mu_{P_j} = 1 - \frac{p_j - p_{j_{min}}}{p_{j_{max}} - p_{j_{min}}}, \quad j = 1, 2, \ldots, N_i \qquad (8.4)$$

The four fuzzy sets of candidate shapes corresponding to the four aforementioned criteria for the design variable x_i are defined as S_{C_i}, S_{W_i}, S_{T_i}, and S_{P_i}. To find the multi-criteria optimum solution, the max-min procedure of Bellman and Zadeh (1970) described in Section 4.2 is used. The solution is

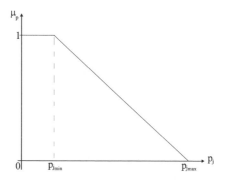

Figure 8.3   Membership function for the minimum perimeter

found by forming a fuzzy set S_{D_i} defined as the intersection of the four fuzzy sets S_{C_i}, S_{W_i}, S_{T_i}, and S_{P_i} as follows:

$$S_{D_j}(\tilde{y}_{ij}) = S_{C_j}(\tilde{y}_{ij}) \cap S_{W_j}(\tilde{y}_{ij}) \cap S_{T_j}(\tilde{y}_{ij}) \cap S_{P_j}(\tilde{y}_{ij}), \quad j = 1, 2, \ldots, N_i$$

$$(8.5)$$

The membership of this intersection fuzzy set S_{D_i} is set equal to the minimum values of the membership functions of the four sets S_{C_i}, S_{W_i}, S_{T_i}, and S_{P_i}:

$$\mu_{D_j}(\tilde{y}_{ij}) = \min\left[\mu_C(\tilde{y}_{ij}); \mu_W(\tilde{y}_{ij}); \mu_T(\tilde{y}_{ij}); \mu_P(\tilde{y}_{ij})\right], \quad j = 1, 2, \ldots, N_i$$

$$(8.6)$$

where $\mu_C(\tilde{y}_{ij})$, $\mu_W(\tilde{y}_{ij})$, $\mu_T(\tilde{y}_{ij})$, and $\mu_P(\tilde{y}_{ij})$ are the fuzzy membership functions for the four criteria: minimizing material cost, minimizing weight, minimizing the number of section types, and minimizing the total surface areas of sections, respectively. The discrete section corresponding to the maximum membership function, $\mu_{D_{i(\text{best})}}$, in the N_i selected candidate shapes represents the best section. This is obtained by using equation (5.11) in Chapter 5. The commercially available discrete shape corresponding to the maximum membership function, $\mu_{D_{i(\text{best})}}$, in the N_i selected candidate shapes represents the best section (Klir and Folger, 1988) and is Pareto optimal.

In obtaining the intersection fuzzy set $S_{D_j}(\tilde{y}_{ij})$ in equation (8.5), it is assumed that all four criteria are of equal importance. However, one of them may be considered as of greater importance than others. In such cases, $S_{D_j}(\tilde{y}_{ij})$ may be expressed as a convex combination of the four criteria with weighting coefficients reflecting their relative importance, as discussed in Section 5.5. Four weighting coefficients w_C, w_W, w_T, and w_P are used for minimizing the material cost, the sectional weight, the number of section types, and the total surface areas of sections, respectively. There is an inverse relationship between the weighting coefficients and their impact on the multi-criteria cost optimization. The modified membership functions are defined as

$$\mu'_{C_i} = w_C \mu_{C_i}$$
$$\mu'_{W_i} = w_W \mu_{W_i}$$
$$\mu'_{T_i} = w_T \mu_{T_i}$$
$$\mu'_{P_i} = w_P \mu_{P_i} \qquad\qquad (8.7)$$

where the sum of the weighting coefficients is equal to 1:

$$w_C + w_W + w_T + w_P = 1 \tag{8.8}$$

8.5 Application to a High-Rise Building Steel Structure

The life cycle cost optimization model presented in this chapter has been applied to a 36-story irregular moment-resisting steel space frame structure with setbacks and cross-bracings discussed in Section 5.9.2 and shown in Figures 5.6 and 5.7. The commercially available shapes and the design constants are the same as in Section 5.9.2.

The design engineer has to make a decision on the relative values of the importance-weighting coefficients for the four membership functions. In this work the following five cases are studied:

Case A $w_C = 0.001$; $w_W = 0.333$; $w_T = 0.333$; $w_P = 0.333$ ($N_i = 6$; minimum material cost design)

Case B $w_C = 0.333$; $w_W = 0.001$; $w_T = 0.333$; $w_P = 0.333$ ($N_i = 6$; minimum weight design)

Case C $w_C = 0.333$; $w_W = 0.333$; $w_T = 0.333$; $w_P = 0.001$ ($N_i = 6$; minimum surface area design)

Case D $w_C = 0.333$; $w_W = 0.333$; $w_T = 0.001$; $w_P = 0.333$ ($N_i = 15$; minimum section type design)

Case E $w_C = 0.25$; $w_W = 0.25$; $w_T = 0.25$; $w_P = 0.25$ ($N_i = 15$; equal preference for the four criteria)

The optimum design solutions obtained by the fuzzy multi-criteria discrete optimization model based on the ASD and LRFD codes for cases A, B, C, D, and E are presented in Tables 8.1 and 8.2. A comparison of the cost values in the two tables leads to the conclusion that optimum design costs based on the LRFD code are consistently less than the corresponding costs based on the ASD code. Compared with the minimum material cost design (case A), the material cost of the minimum surface area design (case C) is 5.2 % more using the ASD code (5.9 % more using the LRFD code). However, the total number of section types in case C is reduced to 38 (36 for LRFD code) from 50 (51 for LRFD code) for case A, resulting in reduced fabrication, erection, storing, and handling costs.

Case E with equal preference for all the criteria appears to yield a good compromise in terms of all the criteria. However, the designer can influence

Table 8.1 Optimum design solutions obtained using the fuzzy discrete multi-criteria cost optimization model based on the ASD code for the 36-story example

Type of study	Material cost of the structure	Weight of the structure	Total number of section types	Total member surface areas
Case A for $N_i = 6$	$6 543 527	15 516.0 kN (3488.3 kips)	50	28 340.5 m^2 (305 054.8 ft^2)
Case B for $N_i = 6$	$6 733 198	15 410.1 kN (3464.5 kips)	58	29 488.9 m^2 (317 416.0 ft^2)
Case C for $N_i = 6$	$6 885 985	15 852.2 kN (3563.9 kips)	38	27 409.9 m^2 (295 037.2 ft^2)
Case D for $N_i = 15$	$7 750 952	18 142.1 kN (4078.7 kips)	20	28 912.9 m^2 (311 215.4 ft^2)
Case E for $N_i = 15$	$7 364 961	17 216.0 kN (3870.5 kips)	24	28 223.3 m^2 (303 793.2 ft^2)

Table 8.2 Optimum design solutions obtained by the fuzzy discrete multi-criteria cost optimization model based on the LRFD code for the 36-story example

Type of study	Material cost of the structure	Weight of the structure	Total number of section types	Total member surface areas
Case A for $N_i = 6$	$6 133 422	14 538.3 kN (3268.5 kips)	51	28 129.8 m^2 (302 786.9 ft^2)
Case B for $N_i = 6$	$6 265 238	14 470.2 kN (3253.2 kips)	60	28 598.6 m^2 (307 832.6 ft^2)
Case C for $N_i = 6$	$6 497 895	14 961.7 kN (3363.7 kips)	36	27 047.1 m^2 (291 132.5 ft^2)
Case D for $N_i = 15$	$7 504 254	17 475.3 kN (3928.8 kips)	23	28 988.8 m^2 (312 032.4 ft^2)
Case E for $N_i = 15$	$6 810 375	15 983.4 kN (3593.4 kips)	25	28 311.2 m^2 (304 739.4 ft^2)

the life-cycle optimum design by using different weighting factors based on his/her experience and considering the prevailing economic and local conditions. The methodology presented in this chapter provides a logical way for the designer to consider the best design for the life cycle of the structure. The methodology can be extended to include additional parameters relevant to the life-cycle design of a structure.

Appendix A

Cross-sectional areas, perimeter, and costs in US dollars for different W-shapes (Nucor, 1999a, 1999b, 1999c) used for axially loaded members

W-shape	Areas (in^2)	Cost in US $ (per foot)	Perimeter ($= 4b_\text{f} + 2d - 2t_\text{w}$) (in)
W6 × 9	2.68	1.35	27.22
W8 × 10	2.96	1.50	31.20
W10 × 12	3.54	1.80	35.20
W6 × 12	3.55	1.80	27.60
W8 × 13	3.84	1.95	31.52
W12 × 14	4.16	2.24	39.30
W10 × 15	4.41	2.25	35.52
W6 × 15	4.43	2.25	35.48
W8 × 15	4.44	2.25	31.79
W12 × 16	4.71	2.40	39.50
W6 × 16	4.74	2.40	28.16
W10 × 17	4.99	2.55	35.78
W8 × 18	5.26	2.70	36.82
W12 × 19	5.57	2.85	39.87
W10 × 19	5.62	2.85	36.06
W6 × 20	5.87	3.00	35.96
W8 × 21	6.16	3.15	37.14
W12 × 22	6.48	3.30	40.22
W14 × 22	6.49	3.30	47.02
W10 × 22	6.49	3.30	42.86
W8 × 24	7.08	3.90	41.35
W6 × 25	7.34	3.75	36.44
W10 × 26	7.61	3.90	43.22

Cost Optimization of Structures: Fuzzy Logic, Genetic Algorithms, and Parallel Computing   H. Adeli and K. C. Sarma   © 2006 John Wiley & Sons, Ltd

(Continued)

W-shape	Areas (in^2)	Cost in US \$ (per foot)	Perimeter ($= 4b_f + 2d - 2t_w$) (in)
W12 × 26	7.65	3.90	49.94
W14 × 26	7.69	3.90	47.41
W8 × 28	8.25	4.20	41.69
W12 × 30	8.79	4.50	50.24
W10 × 30	8.84	4.50	43.58
W14 × 30	8.85	4.50	54.06
W8 × 31	9.13	4.65	47.41
W10 × 33	9.71	4.95	50.72
W14 × 34	10.00	5.10	54.37
W8 × 35	10.30	5.25	47.70
W12 × 35	10.30	5.25	50.64
W14 × 38	11.20	5.70	54.66
W10 × 39	11.50	5.85	51.15
W8 × 40	11.70	6.00	48.06
W12 × 40	11.80	6.00	55.31
W14 × 43	12.60	6.45	58.69
W12 × 45	13.20	6.75	55.63
W10 × 45	13.30	6.75	51.58
W8 × 48	14.10	7.20	48.64
W14 × 48	14.10	7.20	59.02
W10 × 49	14.40	7.35	59.28
W12 × 50	14.70	7.50	55.96
W12 × 53	15.60	7.95	63.41
W14 × 53	15.60	7.95	59.34
W10 × 54	15.80	8.10	59.56
W12 × 58	17.00	8.70	63.70
W8 × 58	17.10	8.70	49.36
W10 × 60	17.60	9.00	59.92
W14 × 61	17.90	9.15	67.01
W12 × 65	19.10	11.05	71.46
W8 × 67	19.70	10.05	49.98
W10 × 68	20.00	10.20	60.38
W14 × 68	20.00	10.20	67.39
W12 × 72	21.10	12.24	71.80
W14 × 74	21.80	11.10	67.72
W10 × 77	22.60	11.55	60.90
W12 × 79	23.20	13.43	72.14
W14 × 82	24.10	12.30	68.12
W12 × 87	25.60	14.79	72.53
W10 × 88	25.90	16.72	61.53
W14 × 90	26.50	15.98	85.24
W12 × 96	28.20	16.32	72.96

W14 × 99	29.10	17.57	85.61
W10 × 100	29.40	19.00	62.20
W12 × 106	31.20	18.02	73.44
W14 × 109	32.00	19.35	86.01
W10 × 112	32.90	21.28	62.87
W12 × 120	35.30	20.40	74.10
W14 × 120	35.30	21.30	86.46
W14 × 132	38.80	23.43	86.93
W12 × 136	39.90	23.12	74.84
W14 × 145	42.70	25.74	90.20
W12 × 152	44.70	27.36	75.60
W14 × 159	46.70	29.42	90.73
W12 × 170	50.00	34.00	76.42
W14 × 176	51.80	32.56	91.38
W12 × 190	55.80	38.00	77.32
W14 × 193	56.80	35.71	92.02
W12 × 210	61.80	42.00	78.22
W14 × 211	62.00	39.04	92.68
W12 × 230	67.70	46.00	79.11
W14 × 233	68.50	43.11	93.50
W12 × 252	74.10	50.40	80.05
W14 × 257	75.60	47.54	94.39
W12 × 279	81.90	55.80	81.20
W14 × 283	83.30	52.35	95.34
W12 × 305	89.60	61.76	82.33
W14 × 311	91.40	58.31	96.34
W12 × 336	98.80	68.04	83.63
W14 × 342	101.00	64.12	97.44
W14 × 370	109.00	69.37	98.43
W14 × 398	117.00	74.62	99.40

Appendix B

Cross-sectional areas, perimeter, and costs in US dollars for different W-shapes (Nucor, 1999a, 1999b, 1999c) used for laterally loaded members

W-shape	Areas (in^2)	Cost in US \$ (per foot)	Perimeter ($= 4b_f + 2d - 2t_w$) (in)
W16 × 26	7.68	3.90	52.88
W16 × 31	9.12	4.65	53.31
W18 × 35	10.30	5.25	58.80
W16 × 36	10.60	5.40	59.07
W16 × 40	11.80	6.00	59.39
W18 × 40	11.80	6.00	59.23
W21 × 44	13.00	6.60	66.62
W16 × 45	13.30	6.75	59.71
W18 × 46	13.50	6.90	59.64
W16 × 50	14.70	7.50	60.04
W18 × 50	14.70	7.50	65.25
W21 × 50	14.70	7.50	67.02
W18 × 55	16.20	8.25	65.56
W24 × 55	16.20	8.25	74.37
W21 × 57	16.70	8.55	67.53
W16 × 57	16.80	8.55	60.48
W18 × 60	17.60	9.00	65.87
W24 × 62	18.20	9.30	74.78
W21 × 62	18.30	9.30	74.14
W18 × 65	19.10	9.75	66.16
W16 × 67	19.70	13.06	72.81
W21 × 68	20.00	10.20	74.48

Cost Optimization of Structures: Fuzzy Logic, Genetic Algorithms, and Parallel Computing   H. Adeli and K. C. Sarma   © 2006 John Wiley & Sons, Ltd

(Continued)

W-shape	Areas (in^2)	Cost in US $ (per foot)	Perimeter ($= 4b_f + 2d - 2t_w$) (in)
W24 × 68	20.10	10.20	82.49
W18 × 71	20.80	10.65	66.49
W21 × 73	21.50	10.95	74.75
W18 × 76	22.30	14.25	79.71
W24 × 76	22.40	11.40	82.92
W16 × 77	22.60	15.01	73.31
W21 × 83	24.30	12.45	75.25
W24 × 84	24.70	12.60	83.34
W27 × 84	24.80	15.33	92.34
W18 × 86	25.30	16.12	80.18
W16 × 89	26.20	18.47	73.91
W30 × 90	26.40	17.32	99.72
W21 × 93	27.30	13.95	75.76
W24 × 94	27.70	14.10	83.85
W27 × 94	27.70	17.15	92.82
W18 × 97	28.50	18.19	80.69
W30 × 99	29.10	19.06	100.06
W16 × 100	29.40	20.75	74.47
W21 × 101	29.80	21.46	90.88
W27 × 102	30.00	17.61	93.21
W24 × 103	30.30	15.45	83.96
W24 × 104	30.60	17.68	98.12
W18 × 106	31.10	19.87	81.08
W30 × 108	31.70	20.79	100.47
W21 × 111	32.70	23.59	91.28
W27 × 114	33.50	20.80	93.72
W30 × 116	34.20	22.33	100.87
W24 × 117	34.40	19.89	98.62
W33 × 118	34.70	22.71	110.54
W18 × 119	35.10	22.31	81.69
W21 × 122	35.90	25.92	91.72
W30 × 124	36.50	26.66	101.23
W27 × 129	37.80	23.54	94.08
W18 × 130	38.20	24.37	81.80
W33 × 130	38.30	25.02	111.06
W24 × 131	38.50	22.27	99.17
W21 × 132	38.80	28.05	92.12
W30 × 132	38.90	28.38	101.57
W36 × 135	39.70	25.99	117.70
W33 × 141	41.60	30.31	111.53
W18 × 143	42.10	26.81	82.40
W27 × 146	42.90	29.56	109.41

W24 × 146	43.00	24.82	99.78
W21 × 147	43.20	31.24	92.72
W30 × 148	43.50	31.82	101.96
W40 × 149	43.80	33.90	122.38
W36 × 150	44.20	28.87	118.35
W33 × 152	44.70	33.82	111.97
W18 × 158	46.30	30.81	83.02
W36 × 160	47.00	32.00	118.72
W27 × 161	47.40	33.81	109.94
W24 × 162	47.70	27.54	100.41
W21 × 166	48.80	36.52	93.14
W40 × 167	49.10	39.24	123.12
W33 × 169	49.50	37.60	112.30
W36 × 170	50.00	34.00	119.10
W30 × 173	50.80	38.49	119.51
W27 × 178	52.30	37.38	110.51
W24 × 176	51.70	37.84	100.54
W36 × 182	53.60	40.49	119.51
W40 × 183	53.70	43.00	123.90
W30 × 191	56.10	42.50	120.10
W24 × 192	56.30	41.28	101.12
W40 × 192	56.50	45.12	145.82
W27 × 194	57.00	40.74	110.86
W36 × 194	57.00	43.16	119.91
W40 × 199	58.40	46.76	139.04
W33 × 201	59.10	44.72	128.91
W24 × 207	60.70	44.50	101.72
W36 × 210	61.80	46.72	120.44
W30 × 211	62.00	46.95	120.75
W40 × 215	63.30	50.52	139.66
W27 × 217	63.80	45.57	111.66
W40 × 221	64.80	51.93	146.76
W33 × 221	65.00	49.17	129.53
W24 × 229	67.20	49.23	102.56
W36 × 230	67.60	51.17	136.16
W36 × 232	68.10	51.62	120.98
W30 × 235	69.00	52.29	121.16
W27 × 235	69.10	49.35	112.26
W33 × 241	70.90	53.62	130.14
W40 × 244	71.70	57.34	147.54
W36 × 245	72.10	54.51	136.60
W40 × 249	73.30	58.51	140.26
W24 × 250	73.50	53.75	103.34
W36 × 256	75.40	56.96	121.80
W27 × 258	75.70	54.18	113.08
W36 × 260	76.50	57.85	137.04
W30 × 261	76.70	58.07	121.98

(Continued)

W-shape	Areas (in^2)	Cost in US $ (per foot)	Perimeter ($= 4b_f + 2d - 2t_w$) (in)
W33 × 263	77.40	58.52	130.54
W40 × 268	78.80	62.98	148.24
W40 × 277	81.30	65.09	141.04
W24 × 279	82.00	59.98	104.36
W36 × 280	82.40	62.30	137.65
W27 × 281	82.60	59.01	113.86
W33 × 291	85.60	64.75	131.38
W30 × 292	85.70	64.97	123.00
W40 × 297	87.40	69.79	141.12
W40 × 298	87.60	70.03	149.04
W36 × 300	88.30	66.75	138.21
W24 × 306	89.80	66.55	105.36
W27 × 307	90.20	65.24	114.68
W33 × 318	93.50	71.55	132.18
W40 × 324	95.30	76.95	141.94
W30 × 326	95.70	73.35	124.00
W36 × 328	96.40	73.80	138.66
W40 × 328	96.40	77.90	149.82
W27 × 336	98.70	71.40	115.66
W30 × 357	104.00	80.32	125.00
W33 × 354	104.00	79.65	133.18
W36 × 359	105.00	80.77	139.48
W40 × 362	106.00	85.97	142.94
W27 × 368	108.00	78.20	116.68
W33 × 387	113.00	87.07	134.18
W30 × 391	114.00	87.97	126.02
W36 × 393	115.60	88.42	140.48
W40 × 397	116.00	94.29	143.94

References

Aarts, E. and Korst, J. (1989) *Simulated Annealing and Boltzmann Machines: A Stochastic Approach to Combinatorial Optimization and Neural Computing*, John Wiley & Sons, Inc., New York.

AASHO (1961) *Standard Specifications for Highway Bridges*, American Association of State Highway Officials, Washington, DC.

AASHTO (1977) *Standard Specifications for Highway Bridges*, American Association of State Highway and Transportation Officials, Washington, DC.

AASHTO (1983) *Standard Specifications for Highway Bridges*, American Association of State Highway and Transportation Officials, Washington, DC.

AASHTO (1992) *Standard Specifications for Highway Bridges*, American Association of State Highway and Transportation Officials, Washington, DC.

Abendroth, R. E. and Salmon, C. G. (1986) Sensitivity study of optimum RC restrained end T-sections. *Journal of Structural Engineering, ASCE*, **112**(8), 1928–1943.

Abuyounes, S. and Adeli, H. (1986) Optimization of steel plate girders via general geometric programming. *Journal of Structural Mechanics*, **14**(4), 501–524.

ACI (1963) *Building Code Requirements for Reinforced Concrete (ACI 318–63)*, American Concrete Institute, Detroit, Michigan.

ACI (1969) *Standard Specifications for the Design and Construction of Reinforced Concrete Chimneys (ACI 307–69)*, American Concrete Institute, Detroit, Michigan.

ACI (1971) *Building Code Requirements for Reinforced Concrete (ACI 318–71)*, American Concrete Institute, Detroit, Michigan.

ACI (1977) *Building Code Requirements for Reinforced Concrete (ACI 318–77)*, American Concrete Institute, Detroit, Michigan.

ACI (1983) *Building Code Requirements for Reinforced Concrete (ACI 318–83)*, American Concrete Institute, Detroit, Michigan.

ACI (1989) *Building Code Requirements for Reinforced Concrete (ACI 318–89)*, American Concrete Institute, Detroit, Michigan.

ACI (1995) *Building Code Requirements for Reinforced Concrete (ACI 318–95)*, American Concrete Institute, Detroit, Michigan.

Cost Optimization of Structures: Fuzzy Logic, Genetic Algorithms, and Parallel Computing H. Adeli and K. C. Sarma © 2006 John Wiley & Sons, Ltd

ACI-ASCE (1974) Tentative recommendations for prestressed concrete flat plates. *ACI Journal*, ACI-ASCE Committee 423, **71**(2), 61–71.

Adamu, A. and Karihaloo, B. L. (1994a) Minimum cost design of RC beams using DCOC. Part I: beams with freely-varying cross-sections. *Structural Optimization*, **7**(4), 237–251.

Adamu, A. and Karihaloo, B. L. (1994b) Minimum cost design of RC beams using DCOC. Part II: beams with uniform cross-sections. *Structural Optimization*, **7**(4), 252–259.

Adamu, A. and Karihaloo, B. L. (1995a) Minimum cost design of RC beams with segmentation using continuum-type optimality criteria. *Structural Optimization*, **9**(3/4), 220–235.

Adamu, A. and Karihaloo, B. L. (1995b) Minimum cost design of RC frames using the DCOC method. Part I: columns under uniaxial bending actions. *Structural Optimization*, **10**(1), 16–32.

Adamu, A. and Karihaloo, B. L. (1995c) Minimum cost design of RC frames using the DCOC method. Part II: columns under biaxial bending actions. *Structural Optimization*, **10**(1), 33–39.

Adamu, A., Karihaloo, B. L. and Rozvany, G. I. N. (1994) Minimum cost design of reinforced concrete beams using continuum-type optimality criteria. *Structural Optimization*, **7**(1/2), 91–102.

Adeli, H. (ed.) (1988) *Expert Systems in Construction and Structural Engineering*, Chapman and Hall, London.

Adeli, H. (ed.) (1990) *Knowledge Engineering – Volume One – Fundamentals*, McGraw-Hill, New York.

Adeli, H. (1992a) *Parallel Processing in Computational Mechanics*, Marcel Dekker, New York.

Adeli, H. (1992b) *Supercomputing in Engineering Analysis*, Marcel Dekker, New York.

Adeli, H. (ed.) (1994) *Advances in Design Optimization*, Chapman and Hall, London.

Adeli, H. (2001) Neural networks in civil engineering – 1989–2000. *Computer-Aided Civil and Infrastructure Engineering*, **16**(2), 126–142.

Adeli, H. and Balasubramanyam, K. V. (1988) *Expert Systems for Structural Design: A New Generation*, Prentice-Hall, Englewood Cliffs, New Jersey.

Adeli, H. and Cheng, N. T.(1993) Integrated genetic algorithm for optimization of space structures. *Journal of Aerospace Engineering, ASCE*, **6**(4), 315–328.

Adeli, H. and Cheng, N. T. (1994a) Augmented Lagrangian genetic algorithm for structural optimization. *Journal of Aerospace Engineering, ASCE*, **7**(1), 104–118.

Adeli, H. and Cheng, N. T. (1994b) Concurrent genetic algorithms for optimization of large structures. *Journal of Aerospace Engineering, ASCE*, **7**(3), 276–296.

Adeli, H. and Ge, Y. (1989) A dynamic programming method for analysis of bridges under multiple moving loads. *International Journal for Numerical Methods in Engineering*, **28**(6), 1265–1282.

Adeli, H. and Hung, S. L. (1995) *Machine Learning – Neural Networks, Genetic Algorithms, and Fuzzy Systems*, John Wiley & Sons, Inc., New York.

Adeli, H. and Kamal, O. (1986) Efficient optimization of space trusses. *Computers and Structures*, **24**(3), 501–511.

Adeli, H. and Kamal, O. (1992) Concurrent optimization of large structures: part II – applications. *Journal of Aerospace Engineering, ASCE*, **5**(1), 91–110.

Adeli, H. and Kamal, O. (1993) *Parallel Processing in Structural Engineering*, Elsevier Applied Science, London.

Adeli, H. and Karim, A. (1997a) Neural dynamics model for optimization of cold-formed steel beams. *Journal of Structural Engineering, ASCE*, **123**(11), 1535–1543.

Adeli, H. and Karim, A. (1997b) Scheduling/cost optimization and neural dynamics model for construction. *Journal of Construction Engineering and Management, ASCE*, **123**(4), 450–458.

Adeli, H. and Kim, H. (2001) Cost optimization of composite floors using the neural dynamics model. *Communications in Numerical Methods in Engineering*, **17**, 771–787.

Adeli, H. and Kumar, S. (1995a) Concurrent structural optimization on massively parallel supercomputers. *Journal of Structural Engineering, ASCE*, **121**(11), 1588–1597.

Adeli, H. and Kumar, S. (1995b) Distributed genetic algorithm for structural optimization. *Journal of Aerospace Engineering, ASCE*, **8**(3), 156–163.

Adeli, H. and Kumar, S. (1999) *Distributed Computer Aided Engineering*, CRC Press, Boca Raton, Florida.

Adeli, H. and Park, H. S. (1995a) Optimization of space structures by neural dynamics model. *Neural Networks*, **8**(5), 769–781.

Adeli, H. and Park, H. S. (1995b) Counterpropagation neural networks in structural engineering. *Journal of Structural Engineering*, **121**(8), 1205–1212.

Adeli, H. and Park, H. S. (1996) Hybrid CPN-neural dynamics model for discrete optimization of steel structures. *Microcomputers in Civil Engineering*, **11**(5), 355–366.

Adeli, H. and Park, H. S. (1998) *Neurocomputing for Design Automation*, CRC Press, Boca Raton, Florida.

Adeli, H. and Saleh, A. (1999) *Control, Optimization, and Smart Structures*, John Wiley & Sons, Inc., New York.

Adeli, H. and Soegiarso, R. (1999) *High-Performance Computing in Structural Engineering*, CRC Press, Boca Raton, Florida.

AISC (1963) *Manual of Steel Construction*, 6th edn, American Institute of Steel Construction, New York.

AISC (1969) *Specification for the Design Fabrication and Erection of Structural Steel for Buildings*, American Institute of Steel Construction, New York.

AISC (1970) *Specification for the Design, Fabrication and Erection of Structural Steel for Buildings, Manual of Steel Construction*, 7th edn, American Institute of Steel Construction, Chicago, Illinois.

AISC (1978) *Manual of Steel Construction*, 8th edn, American Institute of Steel Construction, Chicago, Illinois.

AISC (1986) *Manual of Steel Construction – Load and Resistance Factor Design*, American Institute of Steel Construction, Chicago, Illinois.

AISC (1989) *Manual of Steel Construction – Allowable Stress Design*, American Institute of Steel Construction, Chicago, Illinois.

AISC (1994) *Manual of Steel Construction – Load and Resistance Factor Design*, Vol. I, 2nd edn, American Institute of Steel Construction, Chicago, Illinois.

AISC (1995) *Manual of Steel Construction – Allowable Stress Design*, 9th edn, 2nd revision, American Institute of Steel Construction, Chicago, Illinois.

AISC (2001) *Manual of Steel Construction – Load and Resistance Factor Design*, 3rd edn, American Institute of Steel Construction, Chicago, Illinois.

Al-Salloum Y. A. and Siddiqi G. H. (1994) Cost-optimum design of reinforced concrete beams. *ACI Structural Journal*, **91**(6), 647–655.

Andam, K. A. and Knapton, J. (1980) Optimum cost design of precast concrete framed structures. *Engineering Optimization*, **5**(1), 41–50.

Anderson, D. and Islam, M. A. (1979) Design of multistory frames to sway deflection limitations. *The Structural Engineer*, **57B**(1), 11–17.

Anderson, K. E. and Chong, K. P. (1986) Least cost computer-aided design of steel girders. *Engineering Journal, AISC*, Fourth Quarter, 151–156.

Anderson, E., Bai, Z., Bischof, C., Blackford, S., Demmel, J., Dongarra, J., Du Croz, J., Greenbaum, A., Hammarling, S., McKenney, A. and Sorensen, D. (1999) *LAPACK User's Guide*, 3rd edn, SIAM, http://www.netlib.org/lapack/lug/lapack_lug.html.

Annamalai, N., Lewis, A. D. M. and Goldberg, J. E. (1972) Cost optimization of welded plate girders. *Journal of the Structural Division, ASCE*, **98**(ST10), 2235–2246.

Anonymous (1978) *European Convention for Constructional Steel Works*, European Convention for Construction Steel Works.

Arora, J. S. (1989) *Introduction to Optimum Design*, McGraw-Hill, Inc., New York.

Arora, J. S., Huang, M. W. and Hsieh, C. C. (1994) Methods for optimization of nonlinear problems with discrete variables: a review. *Structural Optimization*, **8**, 69–85.

AS3600 (1988) *Concrete Structures*, Standards Association of Australia, Sydney, Australia.

ASCE (1990) *Design of Steel Transmission Pole Structures*, Task Committee on Updating ASCE Design Guide on Steel Transmission Pole Structures, Structural Division of ASCE, New York.

Balling, R. J. and Yao, X. (1997) Optimization of reinforced concrete frames. *Journal of Structural Engineering, ASCE*, **123**(2), 193–202.

Barr, A. S., Sarin, S. C. and Bishara, A. G. (1989) Procedure for structural optimization. *ACI Structural Journal*, **86**(5), 524–531.

Beightler, C. S. and Phillips, D. H. (1976) *Applied Geometric Programming*, John Wiley and Sons, Inc., New York.

Belegundu, A. D. and Arora, J. S. (1984) A computational study of transformation methods for optimal design. *AIAA Journal*, **22**(4), 535–542.

Bell, L. C. and Brown, D. M. (1976) Guyed tower optimization. *Computers and Structures*, **6**(6), 447–450.

Bellman, R. E. and Zadeh, L. A. (1970) Decision-making in a fuzzy environment. *Management Science*, **17**, B141–164.

Benjamin, J. R. (1968) Probabilistic structural analysis and design. *Journal of Structural Division, ASCE*, **94**(ST7), 1665–1679.

Bhatti, M. A. (1996) Optimum cost design of partially composite steel beams using LRFD. *Engineering Journal, AISC*, First Quarter, 18–29.

BOCA (1999) *The BOCA National Building Code*, 14th edn, Building Officials and Code Administrators International, Inc., Country Club Hills, Illinois.

Box, M. J. (1965) A new method of constrained optimization and comparison with other methods. *The Computer Journal*, **8**(1), 42–52.

Bradley, J., Brown, L. H. and Feeney, M. (1974) Cost optimization in relation to factory structures. *Engineering Optimization*, **1**, 125–138.

Brown, R. H. (1975) Minimum cost selection of one-way slab thickness. *Journal of the Structural Division, ASCE*, **101**(ST12), 2585–2590.

Brown, C. B. and Yao, J. T. P. (1983) Fuzzy sets and strcutural engineering. *Journal of Structural Engineering, ASCE*, **109**(5), 1211–1225.

BS449 (1969) *The Use of Structural Steel in Buildings, Part 2*, British Standards Institution, London.

BSI (1969) *Specification for the Use of Structural Steel in Building, BS 449 : Part 2*, British Standards Institution, London.

BSI (1977) *Specification for the structural use of steelwork: Part 1: Simple construction and continuous construction*, British Standards Institution, London.

Carter, C. J., Murray, T. M. and Thornton, W. A. (2002) Economy in steel. *Modern Steel Construction, AISC*, April.

CEB/FIB (1990), *Comite Euro-International du Beton (CEB) 1990: CEB-FIB Model Code*, Paris, Bulletin d'Information Nos 195 and 196.

Chakrabarty, B. K. (1992a) Models for optimal design of reinforced concrete beams. *Computers and Structures*, **42**(3), 447–451.

Chakrabarty, B. K. (1992b) A model for optimal design of reinforced concrete beam. *Journal of Structural Engineering, ASCE*, **118**(11), 3238–3242.

Chang, C. C., Ger, J. F. and Cheng, F. Y. (1994) Reliability-based optimum design for UBC and non-deterministic seismic spectra. *Journal of Structural Engineering, ASCE*, **120**(1), 139–160.

Cheng, F. Y. and Juang, D. S. (1989) Recursive optimization for seismic steel frames. *Journal of Structural Engineering, ASCE*, **115**(2), 445–466.

Choi, C. and Kwak, H. (1990) Optimum RC member design with predetermined discrete sections. *Journal of Structural Engineering, ASCE*, **116**(10), 2634–2655.

Chou, T. (1977) Optimum reinforced concrete T-beam sections. *Journal of the Structural Division, ASCE*, **103**(ST8), 1605–1617.

Cohn, M. Z. and Lounis, Z. (1993) Optimum limit design of continuous prestressed concrete beams. *Journal of Structural Engineering, ASCE*, **119**(12), 3551–3570.

Cohn, M. Z. and Lounis, Z. (1994) Optimal design of structural concrete bridge systems. *Journal of Structural Engineering, ASCE*, **120**(9), 2653–2674.

Cohn, M. Z. and MacRae, A. J. (1984a) Optimization of structural concrete beams. *Journal of Structural Engineering, ASCE*, **110**(7), 1573–1588.

Cohn, M. Z. and MacRae, A. J. (1984b) Prestressing optimization and its implications for design. *PCI Journal*, **29**(4), 68–83.

Cohon, J. (1978) *Multiobjective Programming and Planning*. Academic Press, New York.

CP110 (1972) *Code of Practice for the Structural Use of Concrete, Part 1*, British Standards Institution, London.

CP110 (1976) *Code of Practice for the Structural Use of Concrete, Part 1*, British Standards Institution, London.

CP110 (1980) *Code of Practice for the Structural Use of Concrete, Part 1*, British Standards Institution, London.

Crawford, A. B. and Jenkins, W. M. (1980) Optimum design of some steel roof structures. *The Structural Engineer*, **58A**(10), 317–325.

CSA (1965) *Steel Structures for Buildings, CSA Standard S16-1965*, Canadian Standards Association, Ottawa, Canada.

CSA (1975) *Steel Structures for Buildings, CSA Standard S16-1975*, Canadian Standards Association, Ottawa, Canada.

CSA (1977) *Code for the Design of Concrete Structures for Buildings, CSA Standard CAN3-A23.3-M77*, Canadian Standard Association, Rexadale, Ontario, Canada.

CSA (1984) *Design of Concrete Structures for Buildings, CSA Standard CAN3-A23.3-M84*, Canadian Standard Association, Rexadale, Ontario, Canada.

De Jong, K. A. (1975) An analysis of the behavior of a class of genetic adaptive systems, PhD dissertation, University of Michigan, Ann Arbor, Michigan.

Dhingra, A. K., Rao, S. S. and Kumar, V. (1992) Nonlinear membership functions in multiobjective fuzzy optimization of mechanical and structural systems. *AIAA Journal*, **30**(1), 251–260.

Dinno, K. S. and Mekha, B. B. (1993) Optimal design of reinforced concrete frames based on inelastic analysis. *Computers and Structures*, **47**(2), 245–252.

Douty, R. (1980) Design of steel connections by math programming. *Journal of the Structural Division, ASCE*, **106**(ST5), 1135–1154.

Dowd, K. and Severance, C. R. (1998) *High Performance Computing*, 2nd edn, O'Reilly & Associates, Inc., Sebastopol, California.

Dumonteil, P. (1992) Simple equations for effective length factors. *Engineering Journal, AISC*, Third Quarter, 111–115.

Enevoldsen, I. and Sorensen, J. D. (1994) Reliability-based optimization in structural engineering. *Structural Safety*, **15**(3), 169–196.

Ennis, D. and Baer, T. (1999) *Using the Origin 2000 for Code Development and Analysis*, http://oscinfo.osc.edu/training/o2k/o2k_new/index.htm.

Erbatur, F., Al Zaid, R. and Dahman, N. A. (1992) Optimization and sensitivity of prestressed concrete beams. *Computers and Structures*, **45**(5/6), 881–886.

Ezeldin, A. S. (1991) Optimum design of reinforced fiber concrete subjected to bending and geometrical constraints. *Computers and Structures*, **41**(5), 1095–1100.

Ezeldin, A. S. and Hsu, C. T. (1992) Optimization of reinforced fibrous concrete beams. *ACI Structural Journal*, **89**(1), 106–114.

Fadaee, M. J. and Grierson, D. E. (1996) Design optimization of 3D reinforced concrete structures. *Structural Optimization*, **12**(2/3), 127–134.

Farkas, J. (1991) Fabrication aspects in the optimum design of welded structures. *Structural Optimization*, **3**(1), 51–58.

Farkas, J. and Jarmai, K. (1994) Minimum cost design of laterally loaded welded rectangular cellular plates. *Structural Optimization*, **8**(4), 262–267.

Farkas, J. and Jarmai, K. (1997) *Analysis and Optimum Design of Metal Structures*, A. A. Balkema, Rotterdam, The Netherlands.

Fereig, S. M. (1996) Economic preliminary design of bridges with prestressed I-girders. *Journal of Bridge Engineering, ASCE*, **1**(1), 18–25.

Fiacco, A. V. and McCormick, G. P. (1968) *Nonlinear Programming: Sequential Unconstrained Minimization Techniques*, John Wiley & Sons, Inc., New York.

Fletcher, R. (1975) An ideal penalty function for constrained optimization. *Institute of Mathematics and Its Applications*, **15**(3), 319–342.

Fletcher, R. and Powell, M. J. D. (1963) A rapidly convergent descent method for minimization. *The Computer Journal*, **6**(2), 163–168.

Frangopol, D. M. (1985) Structural optimization using reliability concepts. *Journal of Structural Engineering, ASCE*, **111**(11), 2288–2301.

Frangopol, D. M. and Moses, F. (1994) Reliability-based structural optimization, in *Advances in the Design Optimization* (ed. H. Adeli), Chapman and Hall, London, pp. 492–570.

Frangopol, D. M., Lin, K.-Y. and Estes, A. C. (1997) Life-cycle cost design of deteriorating structures. *Journal of Structural Engineering, ASCE*, **123**(10), 1390–1401.

Friel, L. L. (1974) Optimum singly reinforced concrete sections. *ACI Journal*, **71**(11), 556–558.

Gill, P. E. and Murray, W. (eds) (1974), *Numerical Methods for Constrained Optimizations*, Academic Press, New York.

Goble, G. G. and DeSantis, P. V. (1966) Optimum design of mixed steel composite girders. *Journal of the Structural Division, ASCE*, **92**(ST6), 25–43.

Goble, G. G. and Lapay, W. S. (1971) Optimum design of prestressed beams. *ACI Journal*, **68**(9), 712–718.

Goldberg, D. E. (1989) *Genetic Algorithms in Search, Optimization, and Machine Learning*, Addision-Wesley, Reading, Massachusetts.

Goldberg, D. E. and Samtani, M. P. (1986) Engineering optimization via genetic algorithm, in Proceedings of 9th Conference on *Electronic Computation*, pp. 471–482.

Golomb, S. W. and Baumert, L. D. (1965) Backtrack programming. *Journal of the Association for Computing Machinery*, **12**(4), 516–524.

Graves Smith, T. R. (1983) *Linear Analysis of Frameworks*, Ellis Horwood Limited, Chichester.

Grefenstette, J. J (1990) 'GENESIS', a general purpose simple GA program in C language.

Gropp, W., Lusk, E. and Skjellum, A. (1994) *Using MPI, Portable Parallel Programming with the Message-Passing Interface*, The MIT Press, Cambridge, Massachusetts.

Grossberg, S. (1982) *Studies of Mind and Brain*, Reidel Press, Boston, Massachusetts.

Gunaratnam, D. J. and Sivakumaran, N. S. (1978) Optimum design of reinforced concrete slabs. *The Structural Engineer*, **56B**(3), 61–67.

Haftka, R. T. and Gurdal, Z. (1992) *Elements of Structural Optimization*, Kluwer Academic Publisher, Boston, Massachusetts.

Hajek, P. and Frangopol, D. M. (1991) Optimum design of shear-wall systems. *Computers and Structures*, **38**(2), 171–184.

Hajela, P. (1990) Genetic search – an approach to the nonconvex optimization problem. *AIAA Journal*, **28**(7), 1205–1210.

Hajela, P., Lee, E. and Cho, H. (1998) Genetic algorithms in topological design of flexural systems. *Computer-Aided Civil and Infrastructure Engineering*, **13**(1), 13–22.

Han, S. H., Adamu, A. and Karihaloo, B. L. (1995) Application of DCOC to optimum prestressed concrete beam design. *Engineering Optimization*, **25**(3), 179–200.

Han, S. H., Adamu, A. and Karihaloo, B. L. (1996) Minimum cost design of multispan partially prestressed concrete T-beams using DCOC. *Structural Optimization*, **12**(2/3), 75–86.

Hecht-Nielsen, R. (1987) Counter propagation networks. *Applied Optics*, **26**(23), 4979–4985.

Hecht-Nielsen, R. (1988) Application of counterpropagation networks. *Neural Networks*, **1**(2), 131–139.

Heinloo, M. and Kaliszky, S. (1981) Optimal design of dynamically loaded rigid-plastic structures. Application: thick-walled concrete tube. *Journal of Structural Mechanics*, **9**(3), 235–251.

Himmelblau, D. M. (1972) *Applied Nonlinear Programming*, McGraw-Hill, Inc., New York.

Holland, J. H. (1975) *Adaptation in Natural and Artificial Systems*, The University of Michigan Press, Ann Arbor, Michigan.

Hooke, R. and Jeeves, T. A. (1961) Direct search solution of numerical and statistical problems. *Journal of the Association for Computing Machinery*, **8**, 212–229.

Huanchun, S. and Zheng, C. (1985) Two-level optimum design of reinforced concrete frames with integer variables. *Engineering Optimization*, **9**(3), 219–232.

Imai, K. (1983) Structural optimization to include material selection. *International Journal for Numerical Methods in Engineering*, **19**(2), 217–235.

IMSL (1980) *IMSL Reference Manual*, 8th edn, International Mathematical and Statistical Library, Houston, Texas.

Itoh, M. (1993) Minimum-weight design of continuous composite girders. *Journal of Structural Engineering*, **119**, 1297–311.

Jendo, S. (1990) Multiobjective optimization, in *Structural Optimization, Volume 2: Mathematical Programming* (eds M. Save and W. Prager), Plenum Press, New York, pp. 311–342.

Jendo, S. and Paczkowski, W. M. (1993) Multicriteria discrete optimization of large scale truss systems. *Structural Optimization*, **6**(4), 238–249.

Jenkins, W. M. (1998) Improving structural design by genetic search. *Computer-Aided Civil and Infrastructure Engineering*, **13**(1), 5–11.

Jones, H. L. (1985) Minimum cost prestressed concrete beam design. *Journal of Structural Engineering, ASCE*, **111**(11), 2464–2478.

Kanagasundaram, S. and Karihaloo, B. L. (1990) Minimum cost design of reinforced concrete structures. *Structural Optimization*, **2**, 173–184.

Kanagasundaram, S. and Karihaloo, B. L. (1991a) Minimum-cost design of reinforced concrete structures. *Computers and Structures*, **41**(6), 1357–1364.

Kanagasundaram, S. and Karihaloo, B. L. (1991b) Minimum-cost reinforced concrete beams and columns. *Computers and Structures*, **41**(3), 509–518.

Karim, A. and Adeli, H. (1999a) Global optimum design of cold-formed steel Z-shape beams. *Practice Periodicals on Structural Design and Construction, ASCE*, **4**(1), 17–20.

Karim, A. and Adeli, H. (1999b) Global optimum design of cold-formed steel hat-shape beams. *Thin Walled Structures*, **35**(4), 275–288.

Karim, A. and Adeli, H. (2000) Global optimum design of cold-formed steel I-shape beams. *Practice Periodicals on Structural Design and Construction, ASCE*, **5**(2), 78–81.

Khaleel, M. A. and Itani, R. Y. (1993) Optimization of partially prestressed concrete girders under multiple strength and serviceability criteria. *Computers and Structures*, **49**(3), 427–438.

Khan, F. R. (1974) New structural system for tall buildings and their scale effects on cities, in Proceedings of Symposium on *Tall Buildings* (ed. F. W. Beaufait), Venderbilt University, Nashville, Tennessee, pp. 99–128.

Kim, H. and Adeli, H. (2001) Discrete cost optimization of composite floors using a floating-point genetic algorithm. *Engineering Optimization*, **33**(4), 485–501.

Kim, J.-H. and Myung, H. (1996) A two-phase evolutionary programming for general constrained optimization problem, in Proceedings of the Fifth Annual Conference on *Evolutionary Programming* (eds L. J. Fogel, P. J. Angeline and T. Back), The MIT Press, Cambridge, Massachusetts, pp. 295–304.

Kim, J.-H. and Myung, H. (1997) Evolutionary programming techniques for constrained optimization problems. *IEEE Transactions on Evolutionary Computation*, **1**(2), 129–140.

Kim, S. H. and Wen, Y. K. (1990) Optimization of structures under stochastic loads. *Structural Safety*, **7**(2–4), 177–190.

Kirk, S. J. and Dell'Isola, A. J. (1995) *Life Cycle Costing for Design Professionals*, 2nd edn, McGraw-Hill, Inc., New York.

Kirsch, U. (1972) Optimum design of prestressed beams. *Computers and Structures*, **2**(4), 573–583.

Kirsch, U. (1973) Optimum design of prestressed plates. *Journal of the Structural Division, ASCE*, **99**(ST6), 1075–1090.

Kirsch, U. (1983) Multilevel optimal design of reinforced concrete structures. *Engineering Optimization*, **6**(4), 207–212.

Kirsch, U. (1993) *Structural Optimization*, Springer-Verlag, New York.

Kirsch, U. and Taye, S. (1989) Structural optimization in design planes. *Computer and Structures*, **31**(6), 913–920.

Klir, G. J. and Folger, T. A. (1988) *Fuzzy sets, Uncertainty, and Information*, Prentice-Hall, Englewood Cliffs, New Jersey.

Kocer, F. Y. and Arora, J. S. (1996) Design of prestressed concrete transmission poles: optimization approach. *Journal of Structural Engineering, ASCE*, **122**(7), 804–814.

Kocer, F. Y. and Arora, J. S. (1997) Standardization of steel pole design using discrete optimization. *Journal of Structural Engineering, ASCE*, **123**(3), 345–349.

Kohonen, T. (1988) *Self-Organization and Associative Memory*, Springer-Verlag, New York.

Koski, J. (1994) Multicriterion structural optimization, in *Advances in Design Optimization* (ed. H. Adeli), Chapman and Hall, London, pp. 194–224.

Koskisto, O. J. and Ellingwood, B. R. (1997) Reliability-based optimization of plant precast concrete structures. *Journal of Structural Engineering, ASCE*, **123**(3), 298–304.

Koumousis, V. K. and Arsenis, S. J. (1998) Genetic algorithms in optimal detailed design of reinforced concrete members. *Computer-Aided Civil and Infrastructure Engineering*, **13**(1), 43–52.

Krishnamoorthy, C. S. and Mosi, D. R. (1981) Optimal design of reinforced concrete frames based on inelastic analysis. *Engineering Optimization*, **5**(3), 151–167.

Lakshmy, T. K. and Bhavikatti, S. S. (1995) Optimum design of trough type folded plate roofs. *Computers and Structures*, **57**(1), 125–130.

Lee, B. S. and Knapton, J. (1974) Optimum cost design of a steel-framed building. *Engineering Optimization*, **1**, 139–153.

Lin, K. and Frangopol, D. M. (1996) Reliability-based optimum design of reinforced concrete girders. *Structural Safety*, **18**(2/3), 239–258.

Lin, C.Y. and Yang, Y.J. (1998) Cluster identification techniques in genetic algorithms for multimodal optimization. *Computer-Aided Civil and Infrastructure Engineering*, **13**(1), 53–62.

Lipson, S. L. and Gwin, L. B. (1977) Discrete sizing of trusses for optimal geometry. *Journal of the Structural Division, ASCE*, **103**(ST5), 1031–1046.

Lipson, S. L. and Russell, A. D. (1971) Cost optimization of structural roof system. *Journal of the Structural Division, ASCE*, **97**(ST8), 2057–2071.

Lorenz, R. F. (1988) Understanding composite beam design methods using LRFD. *Engineering Journal, AISC*, First Quarter, 35–38.

Lounis, Z. and Cohn, M. Z. (1993a) Optimization of precast prestressed concrete bridge girder systems. *PCI Journal*, **38**(4), 60–78.

Lounis, Z. and Cohn, M. Z. (1993b) Multiobjective optimization of prestressed concrete structures. *Journal of Structural Engineering, ASCE*, **119**(3), 794–808.

Lounis, Z. and Cohn, M. Z. (1995a) Computer-aided design of prestressed concrete cellular bridge decks. *Microcomputers in Civil Engineering*, **10**(1), 1–11.

Lounis, Z. and Cohn, M. Z. (1995b) An engineering approach to multicriteria optimization of bridge structures. *Microcomputers in Civil Engineering*, **10**(4), 233–238.

Lounis, Z. and Cohn, M. Z. (1996) An approach to preliminary design of precast preten-sioned concrete girders. *Microcomputers in Civil Engineering*, **11**(6), 381–393.

MacRae, A. J. and Cohn, M. Z. (1987) Optimization of prestressed concrete flat plates. *Journal of Structural Engineering, ASCE*, **113**(5), 943–957.

Majid, K. I., Stojanovski, P. and Saka, M. P. (1980) Minimum cost topological design of steel sway frames. *The Structural Engineer*, **58B**(1), 14–20.

Mau, S. and Sexsmith, R. G. (1972) Minimum expected cost optimization. *Journal of the Structural Division, ASCE*, **98**(ST9), 2043–2058.

Means (1999) *Building Construction Cost Data/RS Means (57th Annual Edition)*, RS Means Company, Kingston, Massachusetts.

Memari, A. M., West, H. H. and Cavalier, T. M. (1991) Optimization of continuous steel plate girder bridges. *Structural Optimization*, **3**(4), 231–239.

Michalewicz, Z. (1995) Heuristic methods for evolutionary computation techniques. *Journal of Heuristics*, **1**(2), 177–206.

Moharrami, H. and Grierson, D. E. (1993) Computer-automated design of reinforced concrete frameworks. *Journal of Structural Engineering, ASCE*, **119**(7), 2036–2058.

Moses, F. (1977) Structural system reliability and optimization. *Computers and Struc-tures*, **7**(2), 283–290.

Moses, F. and Goble, G. (1970) Minimum cost structures by dynamic programming. *Engineering Journal, AISC*, **7**(3), 97–100.

MPI (1995) *The Message Passing Interface (MPI) Standard (Version1.1)*, Message Pass-ing Interface Forum, http://www-unix.mcs.anl.gov/mpi/.

MPI (2000) *The Message Passing Interface (MPI) Standard*, http://www-unix.mcs.anl. gov/mpi/.

Mu, L. and Xianming, W. (1999) Multi-hierarchical durability assessment of existing reinforced concrete structures, in Proceedings of the 8th International Conference on *Durability of Building Materials and Components*, 30 May–3 June 1999, pp. 49–69.

Murray, T. M. (1991) Building floor vibrations. *AISC Engineering Journal*, **28**(3), 102–109.

Naaman, A. E. (1976) Minimum cost versus minimum weight of prestressed slabs. *Journal of the Structural Division, ASCE*, **102**(ST7), 1493–1505.

Naaman, A. E. (1982) Optimum design of prestressed concrete tension members. *Journal of the Structural Division, ASCE*, **108**(ST8), 1722–1738.

Nakamura, T. and Takenaka, Y. (1983) Optimum design of multistory multispan frames for prescribed elastic compliance. *Journal of Structural Mechanics*, **11**(3), 271–295.

Nucor (1999a) *Nucor Steel Price List, Effective Date: January 11, 1999*, Nucor Steel, A Division of Nucor Corporation, Jewel, Texas.

Nucor (1999b) *Nucor-Yamato Price List, February 15, 1999*, Nucor-Yamato Steel Company, Blytheville, Arkansas.

Nucor (1999c) *Nucor Steel, Berkley, Start-up Rolling Forecast Wide Flange Beams, March 24, 1999*, Nucor Steel Company, Berkley.

ODOT (1982) *Bridge Design Regulations*, Ohio Department of Transportation, Columbus, Ohio.

OHBDC (1983) *Ontario Highway Bridge Design Code and Commentary*, 2nd edn, Ministry of Transportation and Communications, Downsview, Ontario, Canada.

OpenMP (1998) *OpenMP C and C++ Application Program Interface*, Version 1.0, October 1998, http://www.openmp.org/specs/mp-documents/cspec.pdf.

OpenMP (2000) *OpenMP Simple, Portable, Scalable SMP Programming*, http://www.openmp.org.

Park, H. S. and Adeli, H. (1997a) Data parallel neural dynamics model for integrated design of steel structures. *Microcomputers in Civil Engineering*, **12**(5), 311–326.

Park, H. S. and Adeli, H. (1997b) Distributed neural dynamics algorithms for optimization of large steel structures. *Journal of Structural Engineering, ASCE*, **123**(7), 880–888.

Park, K. and Grierson, D. E. (1999) Pareto-optimal conceptual design of the structural layout of buildings using a multicriteria genetic algorithm. *Computer-Aided Civil and Infrastructure Engineering*, **14**(3), 163–170.

Park, M. H. and Harik, I. E. (1987) Optimum design of horizontally curved R/C slabs. *Journal of Structural Engineering, ASCE*, **113**(11), 2195–2211.

Paul, H., Das Gupta, N. C. and Yu, C. H. (1990) A geometric programming method for cost-optimal design of a modular floor system. *Engineering Optimization*, **16**(3), 205–220.

Paz, M. (1991) *Structural Dynamics Theory and Computation*, 3rd edn, Van Nostrand Reinhold, New York.

PCI (1983) Guide for design of prestressed concrete poles. Committee on Prestressed Concrete Poles, *PCI Journal*, **28**(3), 22–88.

Pearce, H. T. and Wen, Y. K. (1984) Stochastic combination of load effects. *Journal of Structural Engineering, ASCE*, **110**(7), 1613–1629.

Powell, M. J. D. (1964) Efficient method for finding the minimum of a function of several variables without calculating derivatives. *Computer Journal*, **7**(3), 155–162.

Powell, M. J. D. (1969) A method for nonlinear constraints in minimization problems, in *Optimization* (ed. R. Fletcher), Academic Press, London.

Powell, D. and Skolnick, M. M. (1993) Using genetic algorithms in engineering design optimization with non-linear constraints, in *Proceedings of the Fifth International Conference on Genetic Algorithms* (ed. S. Forrest), Morgan Kaufmann, Los Altos, California, pp. 424–431.

Prakash, A., Agarwala, S. K. and Singh, K. K. (1988) Optimum design of reinforced concrete sections. *Computers and Structures*, **30**(4), 1009–1011.

Punch, W. F., Averill, R. C., Goodman, E. D., Lin, S, C., Ding, Y. and Yip, Y. C. (1994) Optimal design of laminated composite structures using coarse-grain parallel genetic algorithms. *Computer Systems in Engineering*, **5**(4–6), 415–423.

Rajeev, S. and Krishnamoorthy, C. S. (1998) Genetic algorithm-based methodology for design optimization of reinforced concrete frames. *Computer-Aided Civil and Infrastructure Engineering*, **13**(1), 63–74.

Rao, S. S. (1980) Structural optimization by chance constrained programming techniques. *Computers and Structures*, **12**(6), 777–781.

Rao, S. S. (1987a) Description and optimum design of fuzzy mechanical systems. *Journal of Mechanisms, Transmissions and Automation in Design, ASME*, **109**, 126–132.

Rao, S. S. (1987b) Multi-objective optimization of fuzzy structural systems. *International Journal for Numerical Methods in Engineering*, **24**(6), 1157–1171.

Rao, S. S., Sundaraju, K., Prakash, B. G. and Balakrishna, C. (1992a) Fuzzy goal programming approach for structural optimization. *AIAA Journal*, **30**(5), 1425–1432.

Rao, S. S., Sundaraju, K., Prakash, B. G. and Balakrishna, C. (1992b) Multiobjective fuzzy optimization techniques for engineering design. *Computers and Structures*, **42**(1), 37–44.

Rao, S. S., Sundaraju, K., Balakrishna, C. and Prakash, B. G. (1992c) Multiobjective insensitive design of structures. *Computers and Structures*, **45**(2), 349–359.

Ravindra, M. K. and Lind, N. C. (1973) Theory of structural code optimization. *Journal of the Structural Division, ASCE*, **99**(ST7), 1541–1553.

Razani, R. and Goble, G. G. (1966) Optimum design of constant-depth plate girders. *Journal of the Structural Division, ASCE*, **92**(ST2), 253–281.

Ridha, R. A. and Wright, R. N. (1967) Minimum cost design of frames. *Journal of the Structural Division, ASCE*, **93**(ST4), 165–183.

Ringertz, U. T. (1988) On methods for discrete structural optimization. *Engineering Optimization*, **13**(1), 47–64.

Rosenbrock, H. H. (1960) An automatic method for finding the greatest or least value of a function. *The Computer Journal*, **3**(3), 175–184.

Rozvany, G. I. N., Zhou, M. and Sigmund, O. (1994) Optimization of topology, in *Advances in Design Optimization* (ed. H. Adeli), Chapman and Hall, London, pp. 340–399.

Russell, A. D. and Choudhary, K. T. (1980) Cost optimization of buildings. *Journal of the Structural Division, ASCE*, **106**(ST1), 283–300.

Salmon, C. G. and Johnson, J. E. (1996) *Steel Structures Design and Behavior*, HarperCollins College Publishers, New York.

Saouma, V. E. and Murad, R. S. (1984) Partially prestressed concrete beam optimization. *Journal of Structural Engineering, ASCE*, **110**(3), 589–604.

Sarma, K. C. and Adeli, H. (1995) Effect of general sparse matrix algorithm on the optimization of space structures. *AIAA Journal*, **33**(12), 2442–2444.

Sarma, K. C. and Adeli, H. (1996) Sparse matrix algorithm for minimum weight design of large structures. *Engineering Optimization*, **27**(1), 65–85.

Sarma, K. C. and Adeli, H. (1998) Cost optimization of concrete structures. *Journal of Structural Engineering, ASCE*, **124**(5), 570–578.

Sarma, K. C. and Adeli, H. (2000a) Cost optimization of steel structures. *Engineering Optimization*, **32**(6), 777–802.

Sarma, K. C. and Adeli, H. (2000b) Fuzzy genetic algorithm for optimization of steel structures. *Journal of Structural Engineering, ASCE*, **126**(5), 596–604.

Sarma, K. C. and Adeli, H. (2000c) Fuzzy discrete multicriteria cost optimization of steel structures. *Journal of Structural Engineering, ASCE*, **126**(11), 1939–1947.

Sarma, K. C. and Adeli, H. (2001) Bilevel parallel genetic algorithm for optimization of large steel structures. *Computer-Aided Civil and Infrastructure Engineering*, **16**(5), 295–304.

Sarma, K. C. and Adeli, H. (2002) Life-cycle cost optimization of steel structures. *International Journal for Numerical Methods in Engineering*, **55**(12), 1451–1462.

Sarma, K. C. and Adeli, H. (2003) Data parallel fuzzy genetic algorithm for cost optimization of large space steel structures. *International Journal of Space Structures*, **18**(3), 195–205.

Sarma, K. C. and Adeli, H. (2005) Comparative study of optimum design of steel high rise building structures using allowable stress design and load and resistance factor design codes. *Practice Periodical on Structural Design and Construction*, **10**(1), 12–17.

Savic, D. A., Evans, K. E. and Silberhorn, T. (1999) A genetic algorithm-based system for the optimal design of laminates. *Computer-Aided Civil and Infrastructure Engineering*, **14**(3), 187–197.

Saxena, M., Sharma S. P. and Mohan, C. (1987) Cost optimization of Intze tanks on shafts using nonlinear programming. *Engineering Optimization*, **10**(4), 279–288.

Schoenauer, M. and Xanthakis, S. (1993) Constrained GA optimization, in Proceedings of the Fifth International Conference on *Genetic Algorithms* (ed. S. Forrest), Morgan Kaufmann, Los Altos, California, pp. 573–580.

SGI (2000a) *Origin and Onyx2 Theory of Operations Manual*, SGI Technical Publication Library, http://techpubs.sgi.com.

SGI (2000b) *Origin and Onyx2 Programmer's Reference Manual*, SGI Technical Publication Library, http://techpubs.sgi.com.

SGI (2000c) *Origin 2000 Rackmount Owner's Guide*, SGI Technical Publication Library, http://techpubs.sgi.com.

Shih, C. J. and Lai, T. K. (1994) Fuzzy weighting optimization with several objective functions in structural design. *Computer and Structures*, **52**(5), 917–924.

Siddall, J. N. (1972) *Analytical Decision Making in Engineering Design*, Prentice-Hall, Inc., Englewood Cliffs, New Jersey.

Simoes, L. M. C. (1996) Optimization of frames with semi-rigid connections. *Computers and Structures*, **60**(4), 531–539.

Smith, A. E. and Tate, D. M. (1993) Genetic optimization using a penalty function, in Proceedings of the Fifth International Conference on *Genetic Algorithms* (ed. S. Forrest), Morgan Kaufmann, Los Altos, California, pp. 499–505.

Soh, C. K. and Yang, J. (1996) Fuzzy controlled genetic algorithm search for shape optimization. *Journal of Computing in Civil Engineering*, ASCE, **10**(2), 143–150.

Soh, C. K. and Yang, J. (1998) Optimal layout of bridge trusses by genetic algorithms. *Computer Aided Civil and Infrastructure Engineering*, **13**(4), 247–254.

Soltani, M. and Corotis, R. B. (1988) Failure cost design of structural systems. *Structural Safety*, **5**(4), 239–252.

Spires, D. and Arora, J. S. (1990) Optimal design of tall RC-framed tube buildings. *Journal of Structural Engineering*, ASCE, **116**(4), 877–897.

SriVidya, A. and Ranganathan, R. (1995) Reliability based optimal design of reinforced concrete frames. *Computers and Structures*, **57**(4), 651–661.

Surahman, A. and Rojiani, K. B. (1983) Reliability based optimum design of concrete frames. *Journal of Structural Engineering*, ASCE, **109**(3), 741–757.

Syswerda, G. (1989) Uniform crossover in genetic algorithms, in Proceedings of 3rd International Conference on *Genetic Algorithms*, George Mason University, 4–7 June 1989, Morgan Kaufmann Publishers, Inc., pp. 2–9.

Tan, G. H., Thevendran, V., Das Gupta, N. C. and Thambiratnam, D. P. (1993) Design of reinforced concrete cylindrical water tanks for minimum material cost. *Computers and Structures*, **48**(5), 803–810.

Tao, Z., Corotis, R. B. and Ellis, J. H. (1995) Reliability-based structural design with Markov decision process. *Journal of Structural Engineering*, ASCE, **121**(6), 971–980.

Templeman, A. B. (1988) Discrete optimum structural design. *Computers and Structures*, **30**(3), 511–518.

Thakkar, M. C. and Sridhar Rao, J. K. (1974) Optimal design of prestressed concrete pipes using linear programming. *Computers and Structures*, **4**(2), 373–380.

Thierauf, G. and Cai, J. (1998) Parallelization of the evolution strategy for discrete structural optimization problems. *Computer-Aided Civil and Infrastructure Engineering*, **13**(1), 23–30.

Thomas Jr, H. R. and Brown, D. M. (1977) Optimum least-cost design of a truss roof system. *Computers and Structures*, **7**(1), 13–22.

Thurston, D. L. and Sun, R. (1993) Structural optimization of multiple attributes. *Structural Optimization*, **5**(4), 240–249.

Tietz, S. B. (1987) Lifecycle costing and whole-life design. *The Structural Engineer*, **65A**(1), 10–11.

Topping, B. H. V., Sziveri, J., Bahreinejad, A., Leite, J. P. B. and Cheng, B. (1998) Parallel processing, neural networks, and genetic algorithms. *Advances in Engineering Software*, **29**(10), 763–786.

Torres, G. G. B., Brotchie, J. F. and Cornell, C. A. (1966) A program for the optimum design of prestressed concrete highway bridges. *PCI Journal*, **11**(3), 63–71.

UBC (1984) *Uniform Building Code*, International Conference of Building Officials, Whittier, California.

UBC (1988) *Uniform Building Code*, International Conference of Building Officials, Whittier, California.

UBC (1997) *Uniform Building Code, Vol. 2 – Structural Engineering Design Provisions*, International Conference of Building Officials, Whittier, California.

Vanderplaats, G. N. (1984) *Numerical Optimization Techniques for Engineering Design with Applications*, McGraw-Hill, Inc., New York.

Wang, G. and Wang, W. (1985a) Fuzzy optimum design of structures. *Engineering Optimization*, **8**, 291–300.

Wang, G. and Wang, W. (1985b) Fuzzy optimum design of aseismic structures. *Earthquake Engineering and Structural Dynamics*, **13**(6), 827–837.

Wilson, J. L., Wagaman, D. A., Veshosky, C. G., Shi, C. G., Adury, P. and Beidleman, A. R. (1997) Life-cycle engineering of bridges. *Microcomputers in Civil Engineering*, **12**(6), 445–452.

Wolfram, S. (1988) *Mathematica: A System for Doing Mathematics by Computer*, 2nd edn, Addition-Wesley, Redwood City, California.

Xu, L. and Grierson D. E. (1993) Computer-automated design of semirigid steel frameworks. *Journal of Structural Engineering, ASCE*, **119**(6), 1740–1760.

Xu, L., Sherbourne, A. N. and Grierson D. E. (1995) Optimal cost design of semi-rigid, low-rise industrial frames. *Engineering Journal, AISC*, Third Quarter, 87–97.

Yeh, I. (1999) Hybrid genetic algorithms for optimization of truss structures. *Computer-Aided Civil and Infrastructure Engineering*, **14**(3), 199–206.

Yeh, Y. and Hsu, D. (1990) Structural optimization with fuzzy parameters. *Computer and Structures*, **37**(6), 917–924.

Yu, W. W. (2000) *Cold-Formed Steel Design*, John Wiley & Sons, Inc., New York.

Yu, M. and Xu, C. (1994) Multi-objective fuzzy optimization of structures based on generalized fuzzy decision-making. *Computer and Structures*, **53**(2), 411–417.

Yu, C. H., Das Gupta, N. C. and Paul, H. (1986) Optimization of prestressed concrete bridge girders. *Engineering Optimization*, **10**(1), 13–24.

Zadeh, L. A. (1965) Fuzzy sets. *Information and Control*, **8**(3), 338–353.

Zadeh, L. A. (1978) Fuzzy sets as a basis for a theory of possibility. *Fuzzy Sets and Systems*, **1**(1), 3–28.

Zahn, M. C. (1987) The economies of LRFD in composite floor beams. *AISC Engineering Journal*, **24**(2), 87–92.

Zielinski, Z. A., Long, W. and Troitsky, M. S. (1995) Designing reinforced concrete short-tied columns using the optimization technique. *ACI Structural Journal*, **92**(5), 619–626.

Zimmermann, H.-J. (1978) Fuzzy programming and linear programming with several objective functions. *Fuzzy Sets and Systems*, **1**, 45–55.

Index

Allowable Stress Design 82–4, 93, 95,
    107, 119, 134–5, 147–54, 174–5
Arch 17
Augmented Lagrangian 10, 78, 87–92,
    133

Beam 3–7, 10, 19, 21–4
Bilevel parallel fuzzy GA 128, 142–6,
    160–3
Box girder bridge 15–16
Branch and bound method 114
Bridge 14, 15, 19

Cache coherency 125, 127
Cache memory 125, 128
Cache miss 138
Cantilever bridge 15
Column 11
Composite floor 53–75
Compromise programming 113
Concrete
    beam 3–7, 10, 19
    bridge 19
    column 11
    frame 12–14, 19
    pipe 17
    slab 3–5, 7, 19

tensile member 17
transmission pole 11
Concurrent genetic algorithm (GA) 134
Connection Machine 130, 134
Continuum-type optimality criteria
    (CTOC) 10, 13
Counter propagation neural network
    (CPN) 68–9
Course-grained parallelization 128
Cray YMP 134
Crossover 92, 139–40, 160

Deterministic cost optimization 20
Discount rate 170
Distributed computing 130
Distributed genetic algorithm (GA) 134
Distributed memory system 130, 134
Distributed shared memory
    multiprocessor 125, 128
Dynamic load balancing 138
Dynamic programming 22

Effective length factor 107
Elastoplastic analysis 19
Euclidean distance 16
European steel design code 107, 136
Evolutionary computing, *see* Genetic
    algorithm (GA)

Feasible conjugate gradient   5
Feasible directions   17, 19
Fine-grained parallelization   128
Floating point GA   63
Folded plate   17
Fork-and-join paradigm   129
Frame   12–14, 19
Fuzzy augmented Lagrangian GA   87–92
Fuzzy genetic algorithm (GA)   78–99,
    136–55
Fuzzy logic   20, 78–80, 170
Fuzzy logic-based optimization   20, 80
Fuzzy membership function   85–7
Fuzzy set theory   78–81

GA, *see* Genetic algorithm (GA)
General geometric programming   7, 15
Genetic algorithm (GA)   78–99, 133–4,
    136–55
Global optimum   78, 90, 133
Goal programming   113

High-rise   93–5, 118–22, 147–55, 174–5
Hypercube architecture   125

Integer programming   6
IRIX operating system   117

Lagrange multiplier method   4–5, 7
Latency   127, 161
Life-cycle cost   1, 19, 167–9
Life-cycle cost optimization   19–20, 165
Linear programming (LP)   4, 17
Load and Resistance Factor Design
    (LRFD)   135–6, 147–54, 174–5
Loop unrolling   130

Master processor   139–40, 145
Master thread   129
Max-min procedure   81, 105, 108, 113
Membership function   81, 85–7, 110–14
Memory bandwidth   134
Memory hierarchy   127
Message Passing Interface (MPI)   128,
    130, 138–46
Migration scheme   140–2, 146–7,
    159–60

Minimax method   16, 113, 172
Moment-resisting frame   107, 119–22,
    135–6, 147–54
Multi-attribute utility method   113
Multi-criteria cost optimization   101–23
Mutation   92, 139–40, 160

Neural dynamics model of Adeli and
    Park   65–7
Nonuniform memory access (NUMA)
    127

OpenMP   128–9, 136–8
Optimality criteria   10, 13
Origin 2000   125, 137–8, 142

Page migration   127, 138
Page replication   127, 138
Parallel computing   125–32
Parallel processing performance
    optimization   130
Parallel virtual machine (PVM)   138
Pareto optimal   112–14
Partial prestressing   5, 6, 10
Penalty function method   87–8
Piecewise linear programming   14
Piecewise linear regression analysis   83
Piecewise parabolic regression analysis
    83–4
Plate girder   21–4
Post-tensioned prestressed concrete   3
Prestressed concrete beam   4
Prestressed concrete bridge   14–15
Probability of failure   18
Processor farming scheme   138–40,
    158–9
Projected Lagrangian   19

Quadratic fuzzy membership function   85

Racing condition   129
Radius of gyration   83
Reliability-based optimization   18–20
Reliability theory   18
Router   125
Rule-based system   77

Sequential linear programming (LP)
17
Sequential optimization method
113
Sequential quadratic programming
13–14
Sequential unconstrained minimization
technique (SUMT)   12–13,
16–17, 19
SGI Origin 2000   93, 117, 134
Shared memory computer   128
Shear wall   17
Silicon Graphics Inc. 125
Simplex method   16
Single-program multiple data (SPMD)
128
Slave processor   139–40, 145
Slave thread   129
Space truss   82–4, 118–19

Steel
beam   21–4
high-rise structures   118–22, 147–55,
174–5
plate girder   21–4
truss   77, 93, 118–19
Subroutine inlining   131
Synchronization   128

Thread   129, 137
Truss   77, 93, 118–19
Tube-in-tube system   149
Two-point crossover   92

Virtual ring topology   140–2
Voided slab   16

Water tank   16–17
Wind load   19, 95, 122, 150